Guide to
Global Environmental Issues

Guide to
Global Environmental Issues

Terry Lawson Dunn

Fulcrum Publishing
Golden, Colorado

To Walter, my husband and best friend.

Copyright © 1997 Terry Lawson Dunn
Front cover image of the earth from outer space courtesy of NASA.
Book design by Bill Spahr

All rights reserved. No part of this book may be reproduced, stored in a retrieval system, or transmitted in any form or by any means, electronic, mechanical, photocopying, recording, or otherwise, without the prior written permission of the publisher.

Library of Congress Cataloging-in-Publication Data

Dunn, Terry Lawson.
    Guide to global environmental issues / Terry Lawson Dunn.
       p. cm.
    Includes bibliographical references and index.
    ISBN 1-55591-956-1 (pbk.)
    1. Environmental sciences.   2. Global environmental change.
GE105.D86 1997
363.7—dc21   97–38572
                                                                                      CIP

Printed in the United States of America

0 9 8 7 6 5 4 3 2 1

Fulcrum Publishing
350 Indiana Street, Suite 350
Golden, Colorado 80401-5093
(800) 992-2908 • (303) 277-1623

Contents

Editor's Foreword   vii
Introduction   xi

Chapter 1: Climate Change   1
    The Problem   1
    Causes   4
    Consequences   8
    Facts About Climate Change   14
    Climate Change Summary   17

Chapter 2: Ozone Depletion   19
    The Problem   19
    Causes   20
    Consequences   23
    Facts About Ozone Depletion   27
    Ozone Depletion Summary   29

Chapter 3: Tropical Deforestation   31
    The Problem   31
    Causes   35
    Consequences   41
    Facts About Tropical Deforestation   42
    Tropical Deforestation Summary   45

Chapter 4: Biodiversity and Extinction   47
    The Problem   47
    Human Causes of Extinction   52
    Consequences   56
    Facts About Biodiversity and Extinction   57
    Biodiversity and Extinction Summary   60

Chapter 5: Pollution 61
The Problem 61
Causes 61
Facts About Pollution 81
Land Pollution Summary 85
Air Pollution Summary 85
Water Pollution Summary 87

Chapter 6: Global Population Growth 89
The Problem 89
Causes 91
Consequences 93
Slowing Population Growth 94
Facts About Population Growth 95
Global Population Growth Summary 98

Appendix: Organizations 99
Notes 107
Glossary 117
References 121
Index 125

Editor's Foreword

One of the perplexing, and often aggravating, realities of our so-called information age is the scarcity of *good* information—information you can count on, and information you can trust. Take virtually any social issue, and somebody, somewhere, is bent on molding our minds by spoon-feeding us selected "facts" in support of their version of the "truth." Nowhere, perhaps, is this more evident than in the great struggle to preserve environmental quality while maintaining economic development. Lobbyists, legislators, and organizations representing all sides of environmental issues would have us believe one way or another simply because they think it is the correct way to believe. It is "truth." The challenge for society, of course, is to decide whose version of the "truth" is most correct. In this, we depend in great measure on two essential institutions—journalism and education—to provide untainted information we can trust. That is both the premise and promise of Terry Lawson Dunn's *Guide to Global Environmental Issues*.

Despite obvious professional differences, journalists and educators share a special role as society's providers of good information on environmental issues. At least that's how we try to see them. Clearly, they are not perfect because they too are human. They carry their own biases and political agendas to work every day, but, unlike most of us, they try to suppress them until the light is turned off at night. We all know journalists or teachers who have taken advantage of their position to advance personal opinions on the environment, but the fact that such occurrences irritate us is clear evidence that we hold teachers and journalists accountable to an ethical code that we could never shoulder ourselves. Our society depends on both professions to give us the "facts" as nearly as they can determine what those facts are, which turns out to be a big problem because the mind molders provide us with so many "facts" about environmental issues—"facts" that frequently contradict one another. Too often, we are left wondering which "facts" are really "facts."

When I asked Terry to write this book, it was precisely this dilemma I had in mind. Journalists and teachers need *good* information about the environment. They need current, scientifically produced information that has not been filtered through a political agenda or corporate mission. They need access to the "facts" before they have been politicized or embellished for financial gain. That is the kind of information—good information—Terry Lawson Dunn has painstakingly compiled and presented in this unusually useful volume.

Terry has been both a writer and an environmental educator. She understands how difficult it is to find trustworthy information on environmental issues, and she knows how important it is that busy people like journalists and teachers get it in a form that is both easy (in the "user-friendly" sense) and understandable. Since most teachers and journalists do not have biology or chemistry backgrounds, Terry has carefully couched her explanations of the issues in common, everyday language—the same language journalists or teachers might use to tell their own audiences about them. Yet (and this is perhaps the most important feature of this important book) she has done it without reducing the scientific integrity of the facts. Every chapter is meticulously documented, and every statement of fact is carefully traced to the original research that produced it. Readers looking for a rehash of old ideas or an analysis of opinion and philosophy on the environment and sustainable development will be sadly disappointed with this book. It provides current scientific knowledge on contemporary global environmental issues by reporting the findings of research by some of the world's most eminent scientists. The manuscript carefully avoids opinion and interpretation of findings except those given by the scientists themselves. The result is one of the most useful almanacs of environmental information available today.

Users of *Guide to Global Environmental Issues* will find it a quick-fix resource. Carefully organized and indexed, it allows a person in need of "instant" information to find it, read it, understand it, and use it in a matter of seconds or minutes depending on the user's needs. Discussions revolve around descriptions of each issue, what is known about them, and their consequences to society. Keeping with the "quick" idea, Terry has provided at the end of each chapter a list of "quick" and interesting facts related to the issue. These facts provide sound bites for the broadcaster, leads for the writer, and fertile discussion points for the educator.

As one of the volumes in Fulcrum Publishing's growing Environmental Communication Series, *Guide to Global Environmental Issues* fills an important niche. Whereas most of the volumes in the series focus on the "how" of communication, this focuses on the "what." The "what" is critical in the information age. We are bombarded by information every day, and many skillful practitioners knowingly and purposefully select and embellish the "facts" they want us to know. This book, however, is for the professional communicator (whether journalist, teacher, interpretive specialist, or lawyer) who wants accurate information in a readily accessible, easy-to-use format. Good information on the environment—especially information readable and accessible to the nonspecialist—is scarce in the information age. As series editor I am so pleased that Terry Lawson Dunn agreed two years ago to write this book. She has prepared what I believe to be one of the most difficult and yet most useful types of references an author could produce. May the result be a fresh and less ambiguous understanding of the world's environment.

—Sam H. Ham, Ph.D.
College of Forestry, Wildlife and Range Sciences
University of Idaho

Introduction

This book is for journalists, educators, students, and others who need a quick, understandable, and accurate source of information about global environmental issues. If you are faced with an impending deadline to report on an environmental issue, need to prepare a lesson plan, or are hunting for interesting facts for a term paper, this book is for you.

Only recently have people begun to realize that their daily activities have global consequences. The forests we cut, the wetlands we drain, and the exhaust from our cars are not only local or regional problems. The impacts spread to the farthest reaches of the atmosphere, to the most remote ice field, and to the deepest ocean.

Understandably, many people find global environmental issues complicated, confusing, or intimidating. The purpose of this book is to explain scientific information about the global environment to nonscientists from all walks of life—journalists, politicians, community leaders, educators, and everyday citizens—who want to understand global environmental issues, the interrelationships between them, and their impact on our lives.

This guide is organized into six chapters, which cover climate change, ozone depletion, tropical deforestation, biodiversity and extinction, pollution, and global population growth. In Chapter 1, you will learn how a variety of human actions are threatening to change the global climate. Chapter 2 explains how the protective ozone layer that surrounds our planet is being destroyed by chemicals we use every day. Chapter 3 explores why the destruction of tropical rainforests can have a global impact. In Chapter 4, you will learn about the variety of life on earth, and what we stand to lose when plants and animals go extinct. Chapter 5 covers pollution on the land, in the air, and in the world's oceans and freshwater ecosystems. And finally, Chapter 6 discusses how all these global environmental problems are exacerbated by the rapid growth of the world's human population.

Each chapter is divided into two main sections. The first section contains a description of the specific global environmental issue and includes background information, the causes or sources of the problem, and the possible consequences. The second section of each chapter provides interesting facts about the problem—lists intended to give you insight into the magnitude of the issues, rather than the latest breaking news.

The book's appendix refers you to sources for more information. The organizations and agencies listed can provide written materials or direct you to authorities on the subjects. The list of written resources contains books and articles written primarily for nonscientific audiences and should serve as a starting point for the vast selection of books and articles available.

The information in this book has been selected from reliable, research-based sources rather than opinion-based materials. Reputable scientific information is based on a process that gives scientists a systematic way to research ideas and scrutinize results. Although the information that comes out of such a process is not always foolproof, it is the best available when it comes to understanding and responding to these serious environmental problems.

It is my hope that this guide will be a useful and frequently used reference for readers. With access to interesting facts and a solid understanding of these environmental issues, journalists, educators, students, and other readers can help to inform others about these important issues. Beyond this, however, I hope this book will inspire some of you to become more active participants in solving these fascinating and serious problems.

As is the case with most books, this book could not have been completed without the help and support of numerous people. I cannot list them all, but I would like to acknowledge some standouts. I wish to thank those who read early drafts of the manuscript, Lew Nelson, Earnest Ables, Sarah Sheldon, and especially Sam Ham, who not only read early drafts, but who has been a great editor and a wonderful friend during the entire writing process. Thanks to all my family members (my mother, my father, Craig, Mead, and my "new" family, Barbara, Read, John, Ginny, Patten, Judy, and Diane) who have unknowingly been "test cases" when I needed to know how informed "regular" people are about global environmental issues. Thanks to Margaret Pennock, Karen Keagle, and Heather Zehren, my coworkers at World Wildlife Fund, who were especially encouraging when I would drag myself home from the office in order to work on this book. Thanks to little Ryan, my new son, whose impending arrival helped me quickly complete the final draft of this book. And most of all, thanks to my husband, Walter, for all his encouragement, his feedback, and for helping me find the time to work.

Chapter 1
Climate Change

The Problem

The earth's climate is maintained by the interaction between atmospheric gases, the oceans, the earth's orbit around the sun, the polar ice caps, and living organisms. However, the atmosphere (the gaseous layer of air surrounding the earth) and biosphere (the layer of life around earth) are being altered by human activities, and these alterations could change the world's climate.

Normally, sunlight passes through transparent atmospheric gases to the earth's surface. As it passes through the atmosphere, some of the short-wave radiation (sunlight) becomes long-wave radiation (heat). As the earth warms, the heat rises. This heat would go back to space if not for the blanket of atmospheric gases that traps some of it. This process is called the greenhouse effect—a normal phenomenon in which transparent gases allow the sun's radiant energy to pass toward the earth while trapping some of the earth's heat energy on the way back to space. The phrase "greenhouse effect" is often used interchangeably with "global warming" and "climate change" to describe the increasing temperature of the earth's surface. However, the greenhouse effect is a natural phenomenon. "Global warming" and "climate change" refer to possible changes caused by an enhanced greenhouse effect. Without greenhouse gases, the earth's heat loss by long-wave radiation would reduce the temperature of the land surface by 63° F (35° C). The current average temperature of the earth's surface is 62° F (17° C).

The amount of heat radiated from the earth's surface and escaping into space depends on a number of factors. An overriding influence is the normal, long-term variation in the earth's orbit around the sun. The tilt and shape of the earth's orbit change every 22,000 years, 41,000 years, and 100,000 years. Where our planet is oriented at any given time dictates how much solar energy reaches the earth's surface. In turn, changes in the planet's orbit may cause the major ice ages that occur roughly every 100,000 years and the shorter warm and cold spells that occur between.

The albedo effect is another factor that influences the amount of solar radiation that stays to warm the earth's surface. The albedo effect is the relationship between the "shininess" of the earth's surface and the amount of solar radiation that is reflected into space. For example, ice and snow reflect up to 90 percent of the sun's radiation while evergreen forests absorb up to 50 percent of the sun's radiation. Bare soil absorbs up to 70 percent of incoming sunlight. The more radiation that does not return to space, the more the earth heats up.

Greenhouse gases also have a major influence on global temperature. Although these gases occur in only trace amounts in the atmosphere, they have a large impact on the greenhouse effect. Water vapor (clouds, humidity, etc.) is the most abundant of these gases. Second in importance is carbon dioxide (CO_2), which is cycled between the oceans, the atmosphere, and vegetation. The oceans have a stabilizing effect on global temperature. Ocean water absorbs heat, and small marine plants called phytoplankton absorb carbon dioxide. Land vegetation also takes carbon dioxide out of the atmosphere by "inhaling" it and "exhaling" oxygen. When vegetation dies, the carbon is released through decomposition or burning. In this way, the cycle of carbon, passing through the oceans and living organisms, is naturally in equilibrium.

Methane (CH_4) is a greenhouse gas that traps and absorbs twenty times as much radiation per molecule as carbon dioxide, but it is not as abundant in the atmosphere. When methane is pumped out of the earth and burned, its carbon is turned into carbon dioxide, which further increases its impact on the greenhouse effect.

Finally, chlorofluorocarbons (CFCs), nitrous oxides (NO_x), ozone (O_3) in the lower atmosphere, and an array of other, less abundant gases contribute to the greenhouse effect.

Because of the natural balance between greenhouse gases, the oceans, the earth's orbit, the snow and ice of the polar ice caps, and living things, the average global temperature has rarely varied more than 3.6° F (2° C) in the last 10,000 years. At the height of the last ice age, the average global temperature was 9° F (5° C) cooler than it is now.

Since the beginning of the industrial revolution the balance of atmospheric gases has been dramatically altered by human activities.

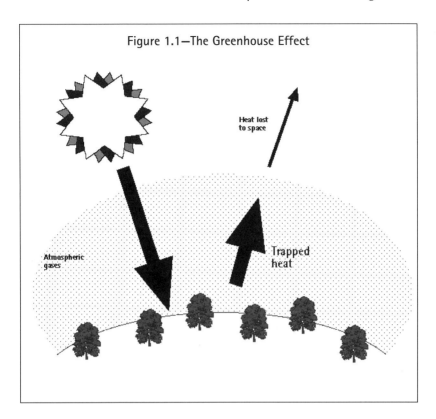

Figure 1.1—The Greenhouse Effect

More carbon dioxide, methane, CFCs, and other greenhouse gases are being pumped into the atmosphere. In a sense, the blanket of heat-trapping gases is thickening, which has the potential to warm the earth.

The theory that global warming could be caused by increasing atmospheric carbon dioxide was first proposed in 1896 by Swedish scientist Svante Arrhenius. It wasn't until 1958 that Charles D. Keeling began to systematically measure the increase of carbon dioxide from the top of Mauna Loa volcano in Hawaii. He recorded an increase in carbon dioxide every year.

Average global temperatures have also been measured over time. Temperatures have increased between 0.54° F (0.3° C) and 1.08° F (0.6° C) this century and are expected to continue to rise by 0.17° F (0.09°C) per decade. The 1980s were the hottest decade ever recorded. After Mount Pinatubo in the Philippines sent a haze of fine ash into the atmosphere in 1991, world temperatures dropped slightly during the following two years. However, the average world temperature in

1994 seemed to continue the trend of the 1980s. In 1994, the average world temperature was only slightly lower than the record high in 1990.

There is some uncertainty whether these measurements indicate that human-induced climate change has begun or whether they are natural climatic fluctuations. There is a growing consensus among scientists, however, that the warming trend indicates that climate change has begun. In one German study released in 1995 by the Max Planck Institute for Meteorology, a sophisticated supercomputer was used to analyze ocean and atmospheric temperature data from the last 1,000 years. The study shows that significant changes have occurred since the beginning of the industrial age. But, scientists still do not agree on the possible rate of change or the consequences we can expect. Some forecasts for the 21st century predict the climate will change 10 to 100 times faster than any changes experienced in recent history.[1]

Causes

Carbon Dioxide

Many scientists estimate that carbon dioxide accounts for roughly 55 percent of the current warming trend. Of the carbon dioxide humans put into the atmosphere, 80 percent comes from burning fossil fuels. The other 20 percent comes from deforestation.

Carbon is released when vegetation is burned or decomposes during deforestation. Because there is three times as much carbon in the living matter of the world than there is in the atmosphere, changes in vegetation cover can dramatically alter the atmosphere. For example, the burning and decomposition of Brazil's forests may account for 6 percent of all carbon dioxide released into the atmosphere.

The level of carbon dioxide in the atmosphere has increased from a pre-industrial level of 280 parts per million (PPM) to 350 PPM in 1990.[2] In 1990, 22.3 billion metric tons of carbon dioxide was released. The United States is responsible for a disproportionate amount of the world's carbon dioxide emissions. With only 5 percent of the world's population, 22 percent of the world's carbon dioxide emissions from fossil fuel burning comes from the United States. Per capita, however, the United States is not first. In 1990, the United Arab Emirates released 33.11 metric tons, the United States released 15.72

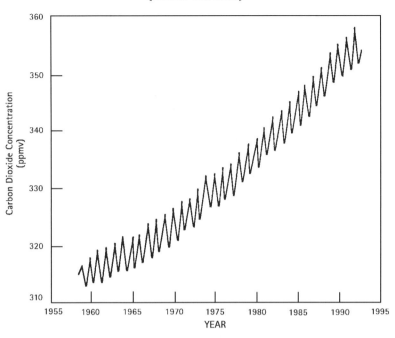

Figure 1.2—Atmospheric Concentration of Carbon Dioxide (Mauna Loa Data)

Source: National Aeronautics and Space Administration

metric tons, and Australia released 15.55 metric tons per capita. Europe and China are rapidly increasing their carbon dioxide emissions as well.

In 1993, world emissions of carbon dioxide decreased slightly from 6.11 billion tons in 1992 to 6.09 billion. A 249-million-ton decrease in emissions in eastern and central Europe was primarily responsible. Emissions from North America, the Middle East, and Africa rose in 1993.

Altogether, the atmospheric concentration of carbon dioxide has increased 30 percent in the last 100 years. Ten percent of that increase has occurred since 1957. To stabilize atmospheric concentrations at today's levels would require an almost 80 percent reduction in world carbon dioxide and other greenhouse gas emissions immediately. Even with drastic reductions, scientists predict the world would still warm about 1° to 4° F (1° to 2° C) and atmospheric levels of carbon dioxide would stay unnaturally high for at least 200 years.[3]

Figure 1.3—Changing Composition of the Earth's Atmosphere

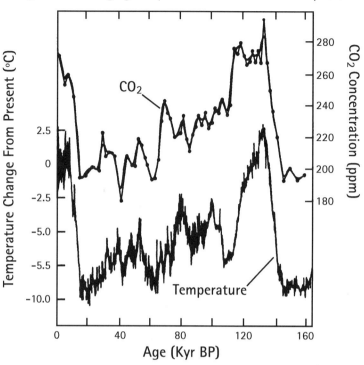

Source: Barnola, J.M.; Raynaud, D.; Korotkevich, Y.S.; and Lorius, C. 1987. Vostok Ice Core Provides 160,000-year Record of Atmospheric CO_2, *Nature*, 329: 408-414

Methane

About 15 percent of the recent warming trend is thought to be caused by increased methane emissions. Scientists think methane concentrations began increasing 300 to 400 years ago and continue to grow 1 to 2 percent a year.[4] They have doubled in just the last 100 years and are expected to double again in seventy years. Even though methane isn't as abundant as carbon dioxide in the atmosphere, methane is more effective at trapping solar radiation and warming the planet.

Methane is emitted naturally from freshwater and marine sediments, burning vegetation, termites digesting deadwood, and bacteria in waterlogged soils such as swamps and marshes. These sources account for 15 to 30 percent of total methane emissions.

Burning coal and other fossil fuels, cultivating rice paddies, disposing of waste in landfills, burning vegetation, leaking natural gas pipelines, and providing termites with more deadwood through deforestation are ways people contribute to methane emissions. In addition, cud-chewing domestic animals, such as cows, goats, sheep, camels, and horses, produce methane in their intestines when they digest food. Methane from these animals, combined with the methane pumped out by rice paddies, accounts for 35 to 40 percent of the methane emissions that can be traced to human activities.[5] The acreage for rice cultivation is increasing as the world population grows, which means more methane will be pumped into the atmosphere.[6]

Burning vegetation releases both carbon dioxide and methane. But even though the volume of methane is only 1 percent of the amount of carbon dioxide emitted from the same burn, that amount of methane adds up to an average of 40 liters a day for every person in the world[7] and 10 to 15 percent of the additional methane on a global basis.[8]

The acceleration of deforestation is also creating more termite food and, therefore, more methane. When a tropical forest is cut down, the termite population can double from 4,000 to 8,000 termites per square yard. Some scientists estimate there may be one-half ton of termites for every person on earth.

Chlorofluorocarbons

Scientists believe chlorofluorocarbons (CFCs) could account for approximately 24 percent of the recent warming trend. CFCs were developed in the early 20th century as a replacement for toxic ammonia in refrigerators. They have also been used in Styrofoam and fire extinguishers, to clean microchips and sterilize hospital equipment, and as a propellant in aerosol cans. There are several kinds of CFCs, each with a slightly different structure. The ones of most concern to climatologists are CFC-11, CFC-12, and CFC-13.

CFCs are efficient heat trappers. One molecule of CFC-11 or CFC-12 can trap as much heat as 10,000 molecules of carbon dioxide. They are also rapidly finding their way into the upper atmosphere where they deplete the ozone layer (the ozone layer protects the earth from the sun's ultraviolet radiation). During the 1980s, CFC-11 and CFC-12 were increasing in the atmosphere by 4 percent a year.[9] The amount of CFCs being released into the atmosphere, however, is currently decreasing as they are phased out in accordance with the Montreal Protocol (see page 24 in Chapter 2).

Ozone, Nitrous Oxides, and Clouds

Ground-level ozone (O_3), from the combustion of fossil fuels, also traps heat. Unfortunately, the excess ozone at ground level can't be used to replenish the ozone layer because of its limited life span.

Nitrous oxides (NO_x), from fertilizers and coal combustion, are also increasing at a rate of 0.3 percent a year. The concentration of nitrous oxides and methane together is expected to double natural levels by the year 2030. Nitrous oxides account for roughly 6 percent of the current warming trend. Nitrous oxides are more efficient than both carbon dioxide and methane at trapping solar radiation.

As mentioned earlier, water vapor is an important greenhouse gas. Like other greenhouse gases, transparent water vapor tends to trap heat. Clouds, on the other hand, are made up of tiny water droplets that reflect solar energy rather than absorb it. At any given time, about one-half of the earth's surface is under cloud cover. Therefore, water in the atmosphere can either enhance or slow global warming depending on the type and thickness of clouds. For this reason, it is difficult to predict the impact of cloud cover on climate change.

Deforestation and Climate Change

Deforestation can influence climate change in three major ways. First, carbon and methane are released when a forest is cut and burned. Second, deadwood increases the food supply for termites, which releases more methane. Finally, if the forest is replaced with rice paddies or cattle, more methane is emitted.

On average, forests can store 100 tons of carbon per acre; tropical rainforests store three to five times more carbon than open, dry forests in temperate areas. Trees and shrubs cover 40 percent of the land on earth (down by one-third since pre-agricultural times) so the potential for future carbon dioxide increases from deforestation is enormous.[10]

Consequences

Models

The consequences of climate change are predicted largely with computer models called General Circulation models. The level of complexity needed to make predictions about climate change is staggering, and scientists are still working with incomplete knowledge of how our planet works.

Even though many of the factors that affect the climate are known, the extent of their influence is uncertain. For instance, the amount of carbon dioxide and heat the oceans can absorb may depend on ocean turbulence, water temperature, and the abundance of phytoplankton. Water vapor and clouds have various effects, depending on their altitude and water content. Ice, snow, and forests all influence the "shininess" of the earth which, in turn, affects the earth's temperature. Sulfur emissions from burning fossil fuels and airborne dust from volcanic eruptions actually cool the earth, but not fast enough to reverse global temperature increases. Even soil moisture can affect the climate; the ability of peat (moisture-absorbing plants found in bogs and swamps) to store carbon is less when soil moisture is lower. Finally, the changes brought on by human-induced climate change could accelerate other effects. One example is the increase in carbon production that would come with faster decomposition accompanying temperature increases.

All of these uncertainties make predicting the regional and local effects of climate change difficult. In addition, each model makes different assumptions, which leads to an array of predictions. The models are tested by using past and present climatic data to predict known global temperatures from the past. The outcomes have been surprisingly accurate.[11]

The scientific predictions of future climatic change from the models do not describe the extremes of possible change, but the middle ground. However, the smallest changes predicted by the models are greater than any climate change experienced in the history of human civilization and perhaps throughout geologic time.[12]

Global Warming Predictions

Given our present scientific knowledge, the predicted temperature change we can expect varies. Estimates range from 2.8° F (1.6° C) to 11.7° F (6.5° C) by the middle of the 21st century. A 7.2° F (4° C) increase is what most models are predicting by the year 2050. Temperatures produced by a 3.6° F (2° C) rise have not been seen on earth for 125,000 years and a 5.4° F (3° C) rise will bring on temperatures not seen for 2 million years.

The temperature changes will be unevenly distributed around the globe. Most of the warming is expected during the Northern Hemisphere winters. The humid tropics are expected to warm less than the global average. Nighttime temperatures are expected to warm more than daytime temperatures. For each 1.8° F (1° C) increase, temperature zones will shift 100 miles. Around the world, the number of hot days may also increase. For example, the number of days over 100° F (38° C) will likely increase from one to twelve in Washington, D.C. Rather than a gradual warming, some scientists think global warming will occur abruptly; a threshold will be crossed and within a relatively short time period we will be living in a warmer world.[13]

Climate and Rainfall

The phrase "climate change" refers to the change in normal climate over decades, not changes from year to year. For this reason, the evidence for climate change is still incomplete.

If the poles heat faster than the equatorial regions, as expected, there will be a smaller difference in temperatures between the two regions. Because it is this temperature difference that creates weather systems and ocean currents, shifts in climatic zones and global weather patterns will likely occur with uneven global warming. In this climate change scenario, a 10 percent increase in rainfall is predicted for the tropics and subtropics, which already receive large amounts of rainfall. In the mid-latitudes (30° to 55°) between northern Mexico and southernmost Alaska, less rain is expected. Generally, more precipitation and cloudiness would occur in the Arctic and Antarctic regions. Seasonal variations are also predicted.

Climatic changes can have many effects. The models show that average worldwide precipitation will increase, but the interior of continents will become drier. Severe storms, such as hurricanes, could increase in number and strength. Flooding and erosion could increase in deserts and other areas that formerly had less rainfall. Fertile areas could turn to desert. It is not yet known whether droughts or floods will increase or decrease overall. Today's water supplies could become undependable, which would affect hydropower systems and groundwater supplies. The consequences might not be bad for everyone, however. Agriculture could boom in higher latitudes of Canada and Siberia, in areas where the soils are suitable for agriculture.

Sea Level Rise

All of the computer models agree that global warming would cause a rise in sea level. Warmer temperatures would expand ocean waters and melt polar ice caps and mountain glaciers.

Some reports already confirm changes to the polar ice caps. According to a 1991 analysis by NASA, arctic ice showed a 2.1 percent decline in surface area between 1978 and 1987. That is equal to a

31,500-square-kilometer loss each year. Norwegian data collected since 1983 show that the polar ice cap is melting 10 percent faster than it is being replaced.[14] And perhaps most dramatic of all, in early 1995, a large portion of the Antarctic Larsen Ice Shelf broke off and began to head toward the warmer waters of the Pacific Ocean. The shelf was 48 miles by 23 miles in size and 650 feet thick. Other ice shelves in the region have also been affected. The Wordie Ice Shelf, which covered 2,000 square kilometers in the 1940s, has lost 75 percent of its mass, and the 900-square-kilometer Price Gustov Shelf has disappeared completely. Since 1930, the average temperature of Antarctica rose by 4° F (2° C). Some scientists predict that melting 10 percent of Antarctica's ice would cause worldwide sea levels to rise between 12 and 30 feet.

Based on 50 years of Australian records and 75 years of New Zealand records, sea levels have already been rising two millimeters per year. If present trends continue, the ocean is expected to rise between one foot (30 centimeters) and five feet (1.5 meters) by the middle of the 21st century.[15]

The consequences of a rising sea could be serious. Coasts, marshes, and swamps would be flooded, leaving millions of people homeless. Many islands would disappear, beaches would erode, and storm damage would worsen. The salinity of coastal rivers and aquifers would increase and marine fisheries would change dramatically.

Some areas would be disproportionately affected. Bangladesh, Egypt, and the Netherlands are examples of areas that are particularly vulnerable to rising seawater. Many of the world's largest cities and half the world's population are in coastal regions, so billions of people could be adversely affected by rising sea levels.

Food Production

Some scientists predict that the global food supply will be unchanged as the earth warms. Rising carbon dioxide in the atmosphere changes the rate that plants grow, respire, use water, and produce seeds. Plant growth and production could increase in areas where soil moisture and nutrients are available. However, many areas where plants may benefit from increasing carbon dioxide may also be affected by new infestations of insect pests that would move with changing climate zones.

Supplies of particular foods could become erratic and each region of the world would be affected differently. In general, models show that agriculture in the lower and middle latitudes would be affected negatively. The higher latitudes would probably benefit from the longer growing season in areas where the soils are productive. Specifically, the midwestern United States, Greece, Italy, France, Germany, and western Australia would produce less food.[16] The hardest-hit regions would be those that are already dry—the Sahal, northern Africa, southern Africa, western Arabia, Southeast Asia, India, Mexico, Central America, the southwestern United States, eastern Brazil, and the Mediterranean countries. Agriculture in Canada, Ukraine, and Siberia would most likely benefit from global warming.[17]

Health

Warmer temperatures mean more heat-related health problems. Heat can worsen cardiovascular, cerebrovascular, and respiratory diseases. The elderly and the sick are most susceptible to heat-related illnesses. Data from New York City show that death rates increase when the temperature rises above 92° F (33° C).[18]

In addition, diseases that are now in the tropics could move northward as the temperature increases. In 1987, a 1.8° F (1° C) average temperature increase in Rwanda led to a 337 percent increase in the malaria rate. Malaria, schistosomiasis, worm infestations, yellow fever, dengue, plague, encephalitis, and dysentery could become diseases of temperate regions in a warmer world. Delayed seasons, higher daily temperatures, and warmer nights are all factors that increase the growth of insect populations.

Effect on Ecosystems

Climate change could have a tremendous impact on ecosystems and biological diversity as well. Every 1.8° F (1° C) rise in the average global temperature could shift temperature zones 100 miles or more.[19] Past changes in worldwide temperatures took thousands of years, and animals and plants had time to adapt or migrate to other areas. As the earth warms more rapidly, animals and plants would have less time to migrate and adapt.

Some ecosystems are more sensitive to changes in temperature. Coral reefs, for example, are susceptible to small changes. They do not grow in water that is too cold, too deep, too fresh, too warm, too rich in nutrients, or too rough. When coral reefs are stressed, they expel their colorful algae, making them appear bleached. Several worldwide coral "bleaching" episodes have been recorded in recent years.

Mangroves and marine ecosystems would certainly be affected by sea level rise, while tundra and permafrost areas would be changed as frozen soil thaws. The southern edges of permafrost in Siberia, Scandinavia, Canada, and Alaska are already melting. In Antarctica, melting ice is releasing long-frozen seeds. According to the British Antarctic Survey, hairgrass plant populations on Galindez Island have increased twenty-five times in just the last 30 years.[20]

Climate change will also affect biological diversity because of species' various tolerances to temperature, rainfall, and humidity. Some species will go extinct if the climate changes beyond what they can adapt to or migrate with. Because trees reproduce slowly and are not mobile, some forests are expected to begin dying. Trees can disperse and migrate on average 60 miles (100 kilometers) per century. Current climate change predictions would require forests to move five to ten times faster.

Animals are better able to migrate. Some species of animals already appear to be moving with changing conditions. Sixteen invertebrate species in Monterey Bay, California, have migrated northward over the past 60 years.

Other species are affected by temperature in surprising ways. A study of painted turtle nests in the Mississippi River showed that when July temperatures averaged 77° F (25° C), all the hatchlings were female. When the average temperature was 70° F (21° C), all the hatchlings were male.[21]

Even though some animal species may be able to migrate or adapt to temperature shifts, the composition of species within ecosystems could be impoverished.

Transportation

In some ways transportation will benefit from climate change. In other ways, it will be hampered. If ice melts and seasons are longer, transportation routes will be more open and ice-free. On the other hand, roads and bridges will need to be modified to deal with rising sea levels.

Unknowns

There are several factors that might accelerate global warming beyond what we are able to predict. For example, nobody knows how much heat and carbon dioxide the oceans can absorb. If the oceans' storage capacity is exceeded, their heat could be released relatively quickly. This could warm the atmosphere abruptly. Likewise, if the storage capacity for carbon dioxide is exceeded, it could also be released into the atmosphere more quickly than normal. The rapid increase in atmospheric carbon dioxide could cause substantial changes in the climate.

Methane can also increase as global temperatures climb. With increasing temperatures, decomposition and methane production in bogs, marshes, and rice paddies will increase. Methane is also trapped in the permafrost of the Arctic; when warmed, this methane would be released.

Recent studies show that clouds absorb more energy from the sun than previously thought. This could change the outcome of climate models, but no one is certain whether the effect of clouds would increase or decrease warming.

The extent to which humans can control the world's environment before future changes are out of our control is unknown. It is possible that the earth will reach a threshold and all our measures to slow climate change will make no difference. Like a row of tumbling dominos, climate change may gain a momentum of its own.

Facts About Climate Change

General

Global temperatures are expected to rise between 2.8° F (1.5° C) and 11.7° F (6.5° C) by the middle of the 21st century.

Scientists believe the average global temperature has increased between 0.54° F (0.3° C) and 1.08° F (0.6° C) in the 20th century.

At present rates, average global temperatures are expected to rise by 0.5° F (0.3° C) per decade.

A temperature increase of 3.6° F (2° C) would produce the highest global temperatures in 125,000 years. A 5.4° F (3° C) rise would produce the highest temperatures in 2 million years.

At the peak of the last ice age, the average global temperature was 9° F (5° C) cooler than it is now.

Nineteen ninety-four was one of the three warmest years ever recorded. The first and second warmest years were also in the 1990s (1990 and 1991).

Annual rate of greenhouse gas increases as of 1993:
   carbon dioxide—0.5 percent per year
   methane—0.9 percent per year
   nitrous oxide—0.3 percent per year
   CFC-11 and CFC-12—4 percent per year

Countries with the highest per capita carbon dioxide emissions:
   United Arab Emirates—33.11 metric tons
   United States—19.74 metric tons
   Canada—15.72 metric tons
   Australia—15.55 metric tons

Trees can disperse and migrate an average of 60 miles (100 kilometers) per century. Current climate change predictions would require forests to move five to ten times faster.

Delayed seasons, higher daily temperatures, and warmer nights all favor increases in insect populations.

United States

With only 5 percent of the world's population, the United States produces 22 percent of the world's carbon dioxide emissions. Most come from burning fossil fuels.

The number of days per year over 100° F (38° C) could increase from 1 to 12 in Washington, D.C.; from 3 to 21 in Omaha, Nebraska; and from 19 to 78 in Dallas, Texas.

A 5.4° F (3° C) rise in temperature in the Great Basin region of the United States could cause the extinction of 44 percent of mammal species and 23 percent of butterfly species.[22]

Rising Sea Levels

Because of expanding ocean waters and melting polar ice caps and glaciers, the sea level is expected to rise by as much as 5 feet (1.5 meters) by the middle of the 21st century.

Arctic ice area decreased by 2.1 percent between 1978 and 1987, which is equal to a 31,500-square-kilometer loss of ice each year.

In early 1995, a large portion of the Antarctic Larsen Ice Shelf broke off and began heading toward the warmer waters of the Pacific Ocean.

The Wordie Ice Shelf in Antarctica, which was 2,000 square kilometers in size in the 1940s, has lost 75 percent of its mass, and the 900-square-kilometer Price Gustov Shelf has disappeared completely.

Ice on Puncak Jaya Kesuma Mountain in Indonesia, which once covered 7.7 square miles, now covers only 1 square mile.

Rising sea levels could put up to 18 percent of Bangladesh under water by the year 2050.[23]

The eastern United States could spend $10 billion to $100 billion to protect their shores if sea levels rise by 1 meter.[24]

Worldwide sea levels rose 3.9 millimeters during both 1993 and 1994, which is twice the rate detected during other years in the 20th century.

Carbon Dioxide (CO_2)

The atmospheric concentration of carbon dioxide has increased 30 percent in 100 years. Even if further emissions were stopped now, the level of carbon dioxide in the atmosphere would remain high for 200 years.

Carbon dioxide levels in the troposphere are higher now than they have been in the last 130,000 years.

Approximately 24 billion metric tons of carbon dioxide is released each year. About four-fifths of that comes from burning fossil fuels and one-fifth comes from felling forests.

Worldwide carbon dioxide emissions decreased slightly from 6.11 billion tons in 1992 to 6.09 billion tons in 1993. There was a 249-million-ton decrease in emissions from eastern and central European countries. Emissions from North America, the Middle East, and Africa rose.

For every gallon of gasoline burned by a car, over 5 pounds of carbon in the form of carbon dioxide is emitted into the atmosphere.

In 1994, scientists reported that pastures established in Brazil, Colombia, and Venezuela in the last 30 years could be storing between 100 million and 500 million tons of carbon per year.

Methane (CH_2)

About one-third of the recent warming trend is thought to be caused by increased methane emissions. Its concentration in the atmosphere has doubled in the last 200 years and at current rates is expected to double again in 70 years.

Sources of methane:
cattle and rice paddies—35 to 40 percent
natural sources—15 to 30 percent
burning vegetation—10 to 15 percent
waste disposal landfills—10 percent[26]

Together, two landfills studied in Maryland were found to emit 300 tons of methane and carbon dioxide annually.

Methane is twenty times more efficient than carbon dioxide at trapping solar radiation and warming the planet.

Some scientists suggest that termites may be responsible for as much as 50 percent of the methane in the atmosphere.

Living tropical forests can hold 4,000 termites per square yard. The population nearly doubles when trees are cut or burned.

If methane emissions were evenly divided among all the world's citizens, each person would be responsible for releasing 40 liters per day.

Chlorofluorocarbons (CFCs)

Some scientists estimate that chlorofluorocarbons may be responsible for up to one-fourth of the global warming.

One molecule of CFC-11 or CFC-12 can trap as much heat as 10,000 molecules of carbon dioxide.

Climate Change Summary

Causes
carbon dioxide (CO2)
methane
chlorofluorocarbons (CFCs)
ozone
nitrous oxides

Consequences

increase in global temperatures

changes in climate patterns (precipitation, flooding, drought, more severe storms)

melting ice caps and glaciers

sea level rise

changes in food production (some supplies could increase, others could decrease)

health (tropical diseases could become more widespread; increases in heat-related health problems and deaths)

ecosystems (shifting temperature zones could affect living conditions of plants and animals)

transportation (longer warm seasons could be beneficial; roads and bridges might need modifications as sea levels rise)

Chapter 2
Ozone Depletion

The Problem

Ozone (O_3) molecules are found from the earth's surface to the upper atmosphere. However, 90 percent of the ozone is concentrated in the stratosphere, which lies 10–30 miles above the earth's surface (the troposphere is just below the stratosphere, 0–10 miles above the earth's surface). This concentration of ozone in the stratosphere is called the ozone layer. The ozone layer is critical to the survival of plants and animals but, at ground level, even small amounts of ozone (one part per million) are toxic to humans.

Until recently, ozone has been produced and destroyed at the same rate. It is naturally produced when sunlight breaks apart oxygen (O_2) molecules in the upper atmosphere. Some of the single oxygen molecules join the pairs of oxygen to create the threesome called ozone. Ozone is broken down when it absorbs ultraviolet

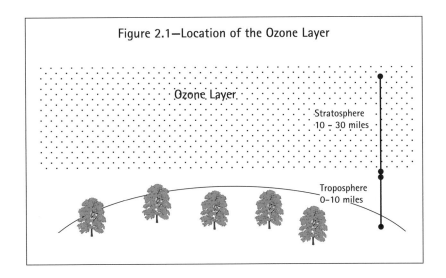

UV-B radiation and by the action of nitrogen oxides (NO_x). Nitrogen oxides naturally come from bacteria in soil and water, and from human and animal waste. Without human interference, these processes keep overall ozone amounts balanced, although the quantity of ozone varies by season, latitude, and solar cycles.

The ozone layer blocks out some of the sun's ultraviolet radiation. In particular, it shields against UV-C radiation, which is lethal in small amounts, and most of the UV-B, which is lethal to many living things (UV-C and UV-B are different wavelengths of ultraviolet light). The UV-B that passes through the ozone layer causes sunburns, skin cancer, and cataracts in people and some animals.

In the late 1970s and early 1980s, the British Antarctic Survey noticed a lower than normal level of ozone during the Antarctic springs (October and November). These initial findings were not reported, but, in 1983, a hole in the ozone layer was discovered. By 1987, the measurements taken over Antarctica showed that the amount of ozone in the region of the hole had dropped by half. In some areas, 95 percent of the ozone was gone. That same year, scientists identified what they suspected was the main cause of the growing hole—chlorofluorocarbons.

Causes

Chlorofluorocarbons

Chlorofluorocarbons (CFCs) were developed in 1930 to replace toxic ammonia used for refrigeration. By 1985, CFCs were a $1.5-billion-a-year market. There are several types of CFCs, including CFC-11, CFC-12, and CFC-13 (each CFC has a slightly different chemical makeup). CFCs are also used in Styrofoam, air conditioners, aerosol sprays (banned in the United States in 1978), fire extinguishers, and pesticides; to fumigate granaries and warehouses; to clean micro chips; and to sterilize hospital equipment. CFCs are sprayed directly into the air in the case of aerosol sprays and fire extinguishers, but they also escape from air conditioners and refrigerator coils and evaporate from liquid cleaners.

There are several other chemicals that destroy ozone. The solvents methyl chloroform and carbon tetrachloride, bromine-containing halons used in fire extinguishers, and methane from natural and industrial sources also destroy ozone. All of these chemicals contribute to global warming as well.

Carbon tetrachloride's ozone depletion potential is greater than all types of CFCs. In Western countries, the use of carbon tetrachloride is restricted because it is toxic and could be a carcinogen. In the United States, western Europe, and Japan, carbon tetrachloride is used in the production of CFCs and as an anticorrosive coating in bridges and ships. In lesser-developed nations, eastern Europe, and the former Soviet Union, it is used as a solvent, dry cleaner, pesticide, and fumigant for grain.

Bromine also has a greater capacity to deplete ozone than CFCs. A single bromine molecule destroys hundreds of times more ozone than one CFC molecule. Scientists predict methyl bromide gas, which is injected into soil to control pests and to help crops grow, will do five times as much damage as CFCs in the next decade. Methyl bromide is used to grow strawberries, tomatoes, eggplant, watermelons, peppers, and squash. If it's banned for use in agriculture, Florida alone could lose 10,000 jobs and $600 million.[1] Methyl bromide is also a by-product of polyester manufacturing.

CFCs don't destroy ozone until the CFC molecule breaks apart. In the troposphere, CFCs stay intact, but when they reach the stratosphere they are broken apart by short-wave ultraviolet radiation.

The chlorine (Cl) atom that breaks off in that process interacts with ozone to form chlorine monoxide (CO) and oxygen (O_2). The chlorine monoxide then combines with free oxygen to form another O_2 and a chlorine atom. What is left is O_2 and a chlorine atom rather than the original ozone (O_3) molecule. Thus, the ozone (O_3) is destroyed by the CFC.

The chlorine atom that remains after the first chain of reactions can destroy up to 100,000 molecules of ozone before being used up in another reaction or falling back to earth. In addition, hydrochloric acid and chlorine nitrate can turn into chlorine monoxide, which destroys ozone.

One reason the ozone hole formed over Antarctica first is that ice crystals in the stratosphere provide a surface on which chlorine molecules can collect. When the sun rises in the Antarctic spring, the ice crystals melt and the chlorine molecules are released. In late November, when the Antarctic spring ends, the molecules stick to the surface of the ice crystals once again.

Another reason the ozone hole may have opened over Antarctica is the polar vortex (a wintertime circular wind pattern). The

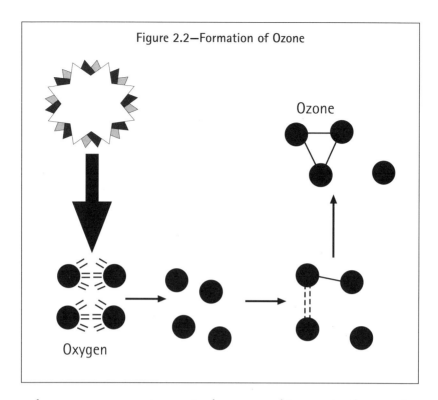

Figure 2.2—Formation of Ozone

polar vortex traps an air mass in the center of Antarctica for months. It also traps CFCs from around the world. When the air over Antarctica warms up, the vortex breaks apart and CFCs are scattered.

The ozone layer is now thinning over the Arctic as well. Between 1979 and 1990, springtime ozone levels over the United States, most of Europe, northern China, and Japan thinned by 8 percent. In early 1992, the ozone layer over the Northern Hemisphere was depleted by 10 to 15 percent, and chlorine and bromine were found in the stratosphere as far south as the Caribbean.[2]

Some types of CFCs can survive in the lower atmosphere for 75 to 120 years. It can take CFCs 10 years to reach the stratosphere once they start rising from ground level. There is also a 10-year supply of CFCs stored in refrigerators, foam, and air conditioners in junkyards. So even if CFCs were banned immediately, there are still enough remaining to destroy the ozone layer for decades.

Unfortunately, the ozone that is created from industrial pollution in the troposphere cannot ascend to the stratosphere. Instead, the rate of ozone production in the stratosphere depends almost exclusively on solar radiation.

Scientists have recently noticed a slowdown in the amount of CFCs released into the atmosphere. In response to lessened demand, companies slowed production of CFCs in the late 1980s. Scientists now expect CFC levels in the atmosphere to peak before the end of the 20th century, then begin to decline.

Consequences

Human Health

As the ozone layer thins, more UV-B radiation will reach the earth's surface. This is expected to cause more sunburns, skin cancers, cataracts, and blindness in people. In addition, increased ultraviolet radiation can suppress human immune systems.

A 2.5 percent reduction in ozone is expected to cause an additional 15,000 melanomas (skin cancers) in the United States each year.[3] Eighty-five percent of exposure to UV-B radiation occurs during daily activities such as driving a car, sitting by a window, and running errands. In 1935, Americans had a one in 1,500 chance of developing a malignant melanoma. In 1991, the odds had increased to one in 150. In the year 2000, the chances may be as high as one in seventy-five.[4] In Australia and Tasmania, the countries closest to the Antarctic ozone hole, melanomas doubled between 1977 and 1987. The Australian government now issues alerts when UV-B radiation is especially intense.

Food Production

The greatest impact of ozone depletion may be on food production. Like people and animals, UV-B radiation can affect plants. Of the plant species tested, two-thirds were damaged in some way when exposed to UV-B radiation.[5] The effects vary from species to species, but photosynthesis and growth are sometimes slowed. An increase in ultraviolet radiation reaching the earth's surface could make it difficult to maintain the current level of food production and even more difficult to provide food for a growing global population.

In the natural world, plants that tolerate UV-B radiation will outcompete plants that don't. The abundance of plant species in a given area is expected to change, which would affect the wildlife that depend on specific habitats.

Earth's Climatic System

When stratospheric ozone blocks incoming UV-B radiation, heat is created. This heat generates strong stratospheric winds, which, in turn, steer weather patterns. Therefore, changes in the ozone layer can change weather and climate patterns. Already, weather patterns seem to be changing over Antarctica.

One possible side effect of ozone depletion is that it may reduce the greenhouse effect. By increasing the amount of sunlight that's reflected back to space, the atmosphere may cool.

Oceans

Plants and animals that live underwater can also be harmed by an increase in UV-B radiation. Under clear conditions, UV-B radiation penetrates up to 20 meters below the water's surface. On the surface, it can kill one-celled plants and krill (small aquatic animals), which are the basis of the entire marine food chain. Plankton and other microscopic organisms also absorb more than half of the world's carbon dioxide emissions. The concentrations of these organisms is highest in Antarctica waters; however, increases in UV-B radiation may have already lowered plankton populations by 6 to 12 percent in the Antarctic oceans.[6]

Ozone depletion may also be a factor in the worldwide decline of other animal populations. The eggs of some frog species are killed or harmed by UV-B, which may explain the worldwide decline of many frog populations. Ozone depletion may also be playing a part in the worldwide die-off of coral reefs.

The Montreal Protocol

In 1988, forty-five nations signed the Montreal Protocol, which called for a 50 percent reduction in CFC production by 1999. However, it did allow developing nations to increase production until they caught up with the basic technologies of industrialized nations. For this reason, the net effect was a 35 percent reduction. In 1989, the European Economic Community (EEC) decided to phase out all CFCs by the year 2000 and reduce them by 85 percent as soon as possible.

Following the example of the EEC, the Montreal Protocol was strengthened in 1990. An updated Montreal Protocol called for a 50 percent reduction of CFCs by 1995, an 85 percent cut by 1997, and a total phaseout by the year 2000. Also, carbon tetrachloride and

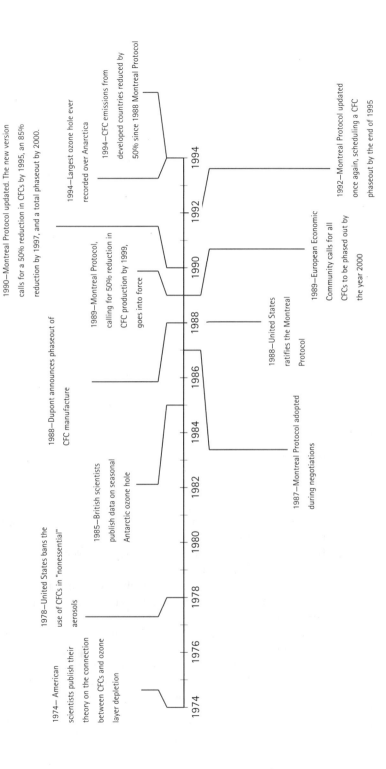

Figure 2.3—Timeline: Ozone Depletion and the Montreal Protocol

methyl chloroform, two other ozone-depleting chemicals, will be eliminated by the years 2000 and 2005 respectively. CFCs used in fire-fighting equipment, aircraft, staffed computer rooms, and for medical purposes are exempt from the phaseout.

In 1992, the Montreal Protocol was updated once again. A 1995 deadline was set for the phasing out of CFCs and other ozone-depleting chemicals. The Montreal Protocol is already showing signs of success. CFC emissions from developed countries have been reduced by 50 percent since the 1988 Montreal Protocol was signed. Unfortunately, the ozone hole continues to grow. In August 1994, the largest hole ever was recorded over Antarctica.

The approaching CFC phaseout has spurred a black market business. In 1994, two men were arrested in Miami for trying to smuggle in 126 tons of CFC-12 worth $1.8 million, without paying a heavy tax imposed by the Environmental Protection Agency (EPA).[7] After 1996, only companies with EPA approval can legally import CFCs.

Figure 2.4—Alternatives to CFCs

Changes that are occurring to prevent CFCs from escaping into the atmosphere:

- CFCs are recaptured from car air conditioners.
- Many micro-electronics manufacturers are using a water-based process for cleaning.
- Foams made of agar, a food additive, are in development, and do not require CFCs to manufacture.
- HFCs (hydrofluorocarbons) are a CFC alternative used in refrigeration; they contain no chlorine, but are still a greenhouse gas.
- HCFCs are also a CFC alternative; they release chlorine, but less than CFCs.
- "Suva," produced by Dupont, is now used in some refrigerator models; Dupont has invested $250 million to develop alternatives to CFCs and has been working on ten new refrigerants.

Facts About Ozone Depletion

The Ozone Layer
Ninety percent of the world's ozone is in the earth's stratosphere.

In the lower atmosphere, ozone is toxic to humans at concentrations of less than one part per million.

The energy it would take to force ground-level ozone into the upper atmosphere is more than twice the world's current power use.

By 1986, ozone was depleted by 50 percent over Antarctica during the polar spring.

Between 1979 and 1990, springtime ozone over the United States, most of Europe, northern China, and Japan thinned by 8 percent.

By 1992, ozone over the Northern Hemisphere was depleted 10 to 15 percent. In August 1994, the ozone hole over the Southern Hemisphere was one and a half times the size of Antarctica.

Ozone Depleters
In 1954, 188 million aerosol cans containing CFCs were sold in the United States; in 1958, 500 million cans were sold; and in 1968, 2.3 billion cans were sold.[8]

By 1985, CFCs were a $1.5-billion-a-year market.

From 1968 to 1975, the concentration of CFCs doubled in the atmosphere. For every chlorine atom released from a CFC molecule, up to 100,000 molecules of ozone are destroyed.

A single polystyrene cup contains more than 1 billion molecules of CFCs. In 1988, food packagers in the United States and England announced plans to stop using containers containing CFCs.

By 1986, 700,000 tons of CFCs were reaching the atmosphere every year.

By 1989, we had already destroyed as much ozone as scientists were predicting we would lose by the year 2050.

CFC-12 can spend up to 120 years in the lower atmosphere before rising to the ozone layer.

CFCs can survive in the stratosphere for a century.

Percentage of CFCs used worldwide each year according to a 1991 report from the Environmental Protection Agency:
Aerosol sprays—3.1 percent
Refrigeration and air-conditioning—15.6 percent
Vehicle air-conditioning—16.2 percent
Foams (packing materials, insulation, etc.)—27.5 percent
Solvents—21.7 percent
Halons—12.0 percent
Miscellaneous—3.8 percent (numbers are rounded)

In 1978, the United States banned the use of CFCs in spray cans.

A 1993 survey of U.S. residents showed that only one in five were aware that CFCs are used in refrigeration and only one in thirty were aware that CFCs are used in air-conditioning.[9]

Over the next decade, methyl bromide, which is used to control pests and grow crops, may do five times more ozone layer damage more than CFCs.

In 1992, Chrysler introduced the Jeep Grand Cherokee, the first U.S. vehicle to use HFC-134a, a CFC alternative. The 1993 line of BMW cars in North America are free of CFCs.

Consequences of Ozone Depletion

Each 1 percent reduction of atmospheric ozone allows 2 percent more UV-B radiation to reach the earth's surface.

More than two-thirds of the plant species tested so far are damaged by excess ultraviolet radiation.[10]

Increases of UV-B radiation over Antarctic oceans have reduced plankton populations 6 to 12 percent.[11]

The United Nations Environment Programme estimates that for every 1 percent loss of stratospheric ozone, there will be an additional 50,000 cases of skin cancer and 100,000 cases of blindness caused by cataracts.

The National Center for Atmospheric Research estimates that future increases in UV-B radiation will increase the skin cancer rates in the Pacific northwestern United States 10 to 20 percent.

The skin cancer rate in Australia doubled from 1977 to 1987.

Eighty-five percent of UV-B exposure occurs during normal daily activities such as running errands, sitting by a window, and driving a car.[12]

In 1935, the chance of an American getting a malignant melanoma was 1 in 1,500. In 1991, it was 1 in 150. And in the year 2000, it may be as high as 1 in 75.[13]

CFC emissions from developed countries dropped by half during the first 7 years of the Montreal Protocol.

Ozone Depletion Summary
Causes
chlorofluorocarbons (CFCs)
methyl chloroform
carbon tetrachloride
methyl bromide

Consequences
increase in ultraviolet radiation
health (more sunburns, skin cancers, cataracts, and suppression of immune system)
food production (more ultraviolet radiation can slow the growth of plants)
climatic systems (possible changes in weather patterns)
oceans (decrease plankton populations)

Chapter 3
Tropical Deforestation

The Problem
Tropical and temperate forests around the world are being cut and burned at an unprecedented rate. Deforestation in the tropics, however, has a greater potential to impact the global environment and is therefore emphasized in this chapter.

The Tropics
Although there are numerous definitions of the "tropics," most people define them as those areas between the latitudes of 22° north and 22° south—better known as the Tropics of Cancer and Capricorn. These are areas that rarely freeze, except in mountainous regions. The mean annual temperature throughout the tropics is 70° F (21° C).

Types of Tropical Forests
Just as there are coniferous forests, broadleaf forests, and deserts in temperate areas, there are a variety of tropical forests. Examples include the Amazon rainforest; deciduous forests that lose their leaves during the dry season; "cloud" forests, which, due to their high elevation, are often enveloped in clouds; and various types of savannas, which are grasslands scattered with trees. Adding to the confusion, many names are given to the same forest type. For example, tropical rainforests are also called wet forests, humid forests, evergreen forests, and lowland rainforests.

Because deforestation is particularly disruptive in tropical rainforests, the focus will be on this forest type. It is important to keep in mind that the following information is generalized. Some applies to cloud forests and deciduous forests as well. In addition, there is a lot of variation in soil types, rainfall, and temperatures that can influence forests in a number of ways.

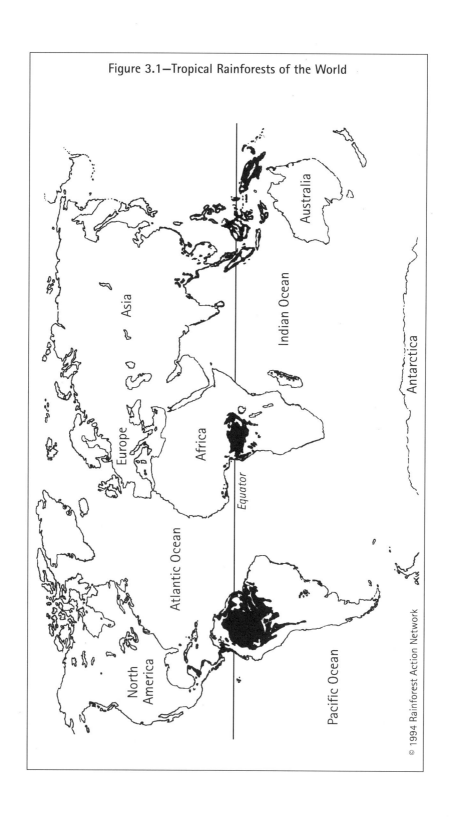

What Is a Tropical Rainforest?

Tropical rainforests are one of the oldest forest types in the world. They have evolved relatively uninterrupted for tens of millions of years.

Latitude, temperature, and rainfall are the main factors that distinguish tropical rainforests. These forests usually lie between 15° north and south of the equator. They have an even year-round temperature and at least 100 inches of rainfall per year. Dry periods in the tropical rainforest rarely exceed one month.

The top (canopy) of a tropical rainforest can reach 150 feet above the forest floor. Some individual trees grow to 250 feet and extend above the forest canopy. Most rainforest animals and plants live in the tangle of branches and vines that make up the canopy.

The forest floor receives only 2 to 5 percent of the sunlight that shines on the canopy. This lack of light leaves the forest floor relatively free of vegetation except in areas where fallen trees create light gaps.

One-half of the world's tropical forests are in Brazil, Zaire, Indonesia, and Peru. In 1990, tropical forest covered 1,756 million hectares (approximately 4,339 million acres). Fifty-two percent (918 million hectares) was in South America and the Caribbean, 30 percent (528 million hectares) was in Africa, and 18 percent (311 million hectares) was in Asia and the Pacific.[1]

Species Diversity

Tropical rainforests have an astounding diversity of plant and animal species. One 750-hectare reserve in Costa Rica contains 320 species of trees, 394 species of birds, 46 species of amphibians, 143 species of butterflies, 104 species of mammals, 42 species of fish, and 76 species of reptiles. Another location in Costa Rica has more bird species than the entire North American continent.[2]

This diversity has resulted in complicated and interesting interrelationships between tropical rainforest species. Many plant species are pollinated by a single species of bee, bat, or hummingbird. The primary source of food for some bird species is insects flushed by the charge of army ants along the forest floor. For Brazil nuts to germinate, the hard seed covering must first be chewed open by a rodent called the agouti. Each species takes advantage of a very specific niche, enabling numerous species to pack into one type of habitat.

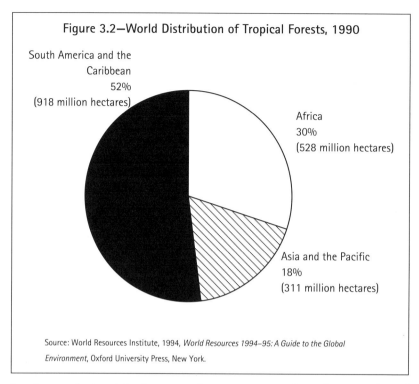

Figure 3.2—World Distribution of Tropical Forests, 1990

South America and the Caribbean 52% (918 million hectares)

Africa 30% (528 million hectares)

Asia and the Pacific 18% (311 million hectares)

Source: World Resources Institute, 1994, *World Resources 1994-95: A Guide to the Global Environment*, Oxford University Press, New York.

Intricate relationships between forest species such as these are found less often in temperate forests. Temperate forest species tend to be generalists that take advantage of a broader range of food sources and habitats.

Many tropical rainforest species also have developed interesting tactics to hide and startle predators. For example, there are insects that look like dead leaves and moths that transform their bodies to look like the head of a viper. But the immense diversity of life in tropical rainforests and lack of extensive research have limited our knowledge about many of the interactions between forest species. It has been said we know less about the canopy of the tropical rainforest than we do about the surface of the moon.

Nutrient Cycles

In general, soils below temperate forests and tropical rainforests are quite different. Soils below tropical rainforests are typically poor. Even though decomposition is much faster in the tropics, most nutrients released by decomposition do not remain in the soil. Instead, they are quickly absorbed by shallow root networks. Nutrients are held in the plants of tropical rainforests rather than in the soils they grow on. Temperate forests, on the other hand, typically grow on more nutrient-rich soils. Nutrients released by decomposition are held in the soil and are only slowly absorbed by the vegetation.

Water Cycle

Rainfall in the tropics is more closely related to the forests than rainfall in temperate regions. Up to 80 percent of the moisture above a tropical forest comes from the forest itself through transpiration (giving off moisture through leaf surfaces) and evaporation. In hot tropical temperatures, plants are continually transpiring water from their leaf surfaces. The water rises to the atmosphere where it forms clouds and eventually returns to the forests in the form of rain. Some locations receive more than 30 feet of rain a year. Without tropical rainforests, regional rainfall patterns would likely change.

All forest types help to slow or prevent soil erosion. They act like a sponge by absorbing rainwater and then slowly discharging it. Because rainwater runoff is controlled, rivers and streams are kept clear and free of silt. A single rainstorm in the tropics can wash away 185 metric tons of topsoil from each deforested hectare.[3]

Albedo Effect

The albedo effect is the earth's reflection of sunlight back into space. Variations on the earth's surface determine how much light is reflected. For example, snow is highly reflective, deserts are somewhat less reflective, and evergreen forests absorb light.

The albedo effect is one of the ways the earth's climate is controlled. It has an impact on rainfall, convection patterns (the upward movement of warm air), and wind currents. If large temperate or tropical forests are cleared, the amount of light reflected back into space increases. This is especially important if deforestation occurs in the tropics, since the tropical regions introduce a lot of energy into the circulation of the global atmosphere and therefore have a large influence on global weather patterns.[4]

Causes

Deforestation is the term used to describe a permanent change in land use. The United Nations defines deforestation as the permanent removal of canopy cover of trees to less than 10 percent of the original amount. Deforestation occurs when a forest is converted into agricultural or urban land uses. If the forest is left to regrow, it is not considered deforestation. Still, the forests that regrow do not always contain the original species diversity.

The average global deforestation rate in the tropics between 1980 and 1990 was 0.8 percent per year, or 15.4 million hectares. Brazil and Indonesia together account for 45 percent of the total tropical rainforest loss.[5] In the Brazilian Amazon, the area of tropical forest that is degraded or fragmented may actually be greater than the amount of land that is deforested completely. Therefore, habitat loss and the loss of biological diversity may be higher than estimates using deforestation data alone. The deforestation rate in Brazil slowed between 1988 and 1990 from 3.5 million hectares to 1.3 million hectares per year, partly due to international pressure to slow deforestation, the establishment of new reserves, and changes in government policies.[6] In other parts of the world, the rate of deforestation is quickening. In Myanmar (formerly Burma), the deforestation rate increased from 105,000 hectares in 1980 to 677,000 in 1990.[7]

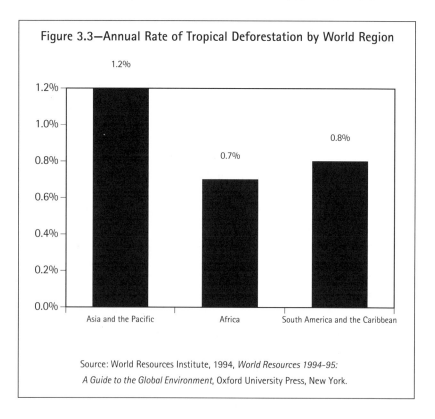

Figure 3.3—Annual Rate of Tropical Deforestation by World Region

Source: World Resources Institute, 1994, *World Resources 1994-95: A Guide to the Global Environment*, Oxford University Press, New York.

By region, Asia and the Pacific are losing tropical forests at a rate of 1.2 percent (3.9 million hectares) per year. Africa's deforestation rate is 0.7 percent (4.1 million hectares) annually. And South America and the Caribbean are losing forests at a rate of 0.8 percent (7.4 million hectares) per year.[8] Overall, Latin America has lost 37 percent of its original tropical moist forest. Asia has lost 42 percent and Africa 52 percent.[9]

Deforestation is occurring for many reasons, including road building, government policies, and some types of agriculture, timber harvesting, fuelwood gathering, and livestock grazing. The reasons for and methods of land clearing are not always new, but even time-honored techniques become unsustainable when forests have to support a growing human population.

Roads

The worst effect of road development in the tropics is that it opens up new areas to subsistence farmers, ranchers, loggers, hunters, and miners. In this way, roads pave the way to many unsustainable forest uses. Without roads, people would not have easy access to forests.

Roads also cause soil erosion and sedimentation of streams and rivers. Humidity, temperature, light, and wind change along the forest edges created by roads. These changes, and the road itself, fragment wildlife habitat.

Agriculture

Across the globe, agriculture for subsistence and cash is a major cause of deforestation. Interestingly, in many temperate areas, the pace of deforestation has slowed. Abandoned farmlands are turning back into forests naturally and through reforestation and fire suppression. Still, virgin forests such as the Douglas fir, western hemlock, giant sequoia, and redwood forests of the western United States and the loblolly pine forests of the Southeast are gradually becoming more rare.

In tropical regions, however, forests are being cleared faster than they are growing back. People living in developing nations often use ancestral agricultural methods. One of these methods is shifting cultivation (also called slash-and-burn agriculture), in which a small patch of forest is cut and burned and crops are planted for family use. Fire is used to clear the land of downed trees, bushes, and grasses to flush game and to reduce snakes, pests, and weeds. After several years, crop yields decrease, the plot is abandoned, the forest eventually regrows, and the process begins again in a new area. Many forests that we consider untouched have, in fact, been previously altered by shifting cultivation.

One difference today is human population growth. Fields carved out of tropical rainforests for agriculture don't have time to recover because more people are using the land. Fields are recut, burned, sprayed with chemicals to control weeds, or used for grazing before they are allowed to recover. The sizes of the fields also have increased as more food is grown to sell, rather than to meet the needs of a single family. Forests are less likely to recover their original complexity when a large area is cleared because some species go extinct, seed sources are farther away, soils change, and even local weather patterns can be altered.

Timber Harvesting

When timber harvesting is unsustainable and forests are unable to regrow, the area is considered deforested. Extracting timber is a major cause of deforestation, particularly in Southeast Asia, Africa, and some parts of the temperate world. Since World War II, incomes and populations have increased in Japan, western Europe, and the United States. The demand for hardwoods has also increased. Simultaneously, the supply of hardwoods has shrunk in parts of the temperate world. In addition, new machinery has made logging and transportation more efficient and new pulping techniques have made a larger variety of hardwoods useful. These factors have led to increased logging in tropical forests.

Worldwide, 5.9 million hectares are logged each year in the tropics. Of that, 4.9 million hectares have never been logged before. For every six hectares logged, only one hectare is replanted as a plantation.[10]

Southeast Asia is the source of three-fourths of exported tropical timber. Africa accounts for most of the rest. The increase in Asian deforestation during the 1980s is partially due to economic growth (and growth in demand for timber) in Thailand and Vietnam as well as investment by Japan and other countries. Asia has designated 49 percent of its forest for logging, but it also has a large percentage (9 percent) of tropical forest set aside for wildlife protection. In contrast, Africa and Latin America only have 3 percent of their forests under protection.[11] More than 70 percent of tropical hardwoods are exported from Indonesia, Malaysia, the Philippines, Papua New Guinea, Brazil, and the Ivory Coast.

Most hardwoods going to Europe come from the tropical lowland and deciduous forests of west and west-central Africa. Latin America's timber is used locally and exported, but Latin America is not a major exporter of tropical hardwoods. Japan receives in excess of 53 percent of the tropical hardwood trade. Europe receives 32 percent and the United States receives 15 percent.[12]

In tropical rainforests, valuable timber trees are often scattered throughout the forest rather than being densely concentrated, as they usually are in temperate forests. It is common to log only a single tree per hectare in the Amazon. However, up to half of the surrounding trees can be killed or damaged when one tree is cut down and removed.

Other forest uses may be more economically valuable in the long run. One study from the Peruvian Andes showed that rubber tapping, extracting fruit, gathering medicinal plants and oils, and agroforestry produce seven times more income per hectare than logging over a 50-year time span.[13]

There is some more bright news about deforestation, however. India now plants four times the number of trees it removes each year.[14] China announced plans in mid-1995 to reforest deserts that had been created more than 1,000 years ago by logging.[15] And Cambodia has banned export of logs and timber.[16]

Fuelwood Gathering

Half of the wood extracted from the world's forests and woodlands each year is used for fuel (the rest is used for wood products). In developing countries, using trees for fuel is an important cause of deforestation. Sub-Saharan African and Indian forests are particularly affected by fuelwood gatherers. Again, roads play an important role in providing forest access to fuelwood gatherers.

Almost half of the world's population depends on wood for heating, cooking, and light. Unfortunately, technologies used for these purposes are not always energy efficient and many forests are cut down at an unsustainable rate. In some developing countries, forests are not recovering fast enough to provide people with fuelwood to meet their daily energy needs. Worldwide, 1.3 billion people are using fuelwood faster than it is being replaced.[17]

Livestock Grazing

Abandoned farmland is often used for livestock grazing in the tropics. Sometimes tropical forests are cleared specifically for cattle ranching.

Clearing forests to create pastures for cattle has been a dominant cause of deforestation in Central America over the past several decades. The intensive grazing and soil compaction caused by cattle ranching often prevent forest regrowth.

A portion of the beef grown in developing countries is exported to developed countries and used in the fast-food industry and in processed meat. Tropical beef has also been used in pet foods.

Government Incentives

Many governments unintentionally subsidize deforestation. The pressure to pay off national debts and to feed and employ growing populations forces governments to promote development of forested lands. Selling timber and other forest products is one way to generate capital and employ rural populations.

Although government policies have changed somewhat, past policies in Brazil offer a dramatic example of how governments can promote deforestation. Roads were built deep into the Amazon to promote settlement at a time when opportunities for Brazil's poor were diminishing in the cities and farmlands of the northeast. Subsidized agricultural loans helped settlers finance farms and ranches. If settlers made "improvements" on the land, they eventually gained title to it. In fact, livestock grazing would not be profitable in the Brazilian Amazon without government subsidies. Unfortunately, the land is not able to sustain the needs of the settlers and people are forced to move deeper into the forests. By 1990, half of the deforested areas of the Brazilian Amazon were abandoned.[18]

Miscellaneous Causes

Forests around the world are also destroyed through more localized causes such as urbanization, mining, war, and flooding caused by dam building. These causes are not as universal as agriculture and timber harvesting, but the local impacts can be devastating. One example of forest destruction during war is the use of chemical defoliants during the Vietnam War, which has had a devastating effect on the forests of that region. In Colombia, poppy and coca cultivation may destroy up to 33 percent of remaining forests by the year 2000. European and

North American forests are also affected by "Waldsterben," or forest death. The causes of forest death are uncertain, but may include pollution combined with disease and climate stress. (See more on Waldsterben in Chapter 5.)

Consequences

Extinction

For some, the most obvious consequence of deforestation is extinction of species. Deforestation destroys habitat for many animals and plants. Some scientists estimate that up to 100 species are becoming extinct every day. Many of the species are not yet known to science. No one is certain how many extinctions an ecosystem can withstand before it ceases to function normally.

Deforestation also impacts migratory animals, such as birds that migrate between North America and Latin America. One-fourth of the birds that live in the United States also live in the tropics part of the year. Deforestation in the tropics may be changing the habitats of these birds. Some scientists have noticed a 1 to 4 percent decrease in the number of some bird species returning to the United States each spring.[19]

Climate and Weather Patterns

Whenever vegetation is burned, carbon dioxide, carbon monoxide, and methane are released into the atmosphere. Rotting vegetation releases carbon and attracts methane-producing termites that feed on deadwood. These chemicals are greenhouse gases and could contribute to climate change.

Large-scale tropical deforestation breaks the water cycle that regulates weather patterns. Since much of the rainfall in the tropics comes from plant transpiration and evaporation, deforestation decreases the total amount of rainfall in certain regions.

Finally, deforestation may alter the albedo effect. With fewer evergreen forests, more sunlight is reflected back into space. Because the albedo effect impacts convection patterns, wind currents, and rainfall patterns, changes in the albedo effect would be expected to change the weather.[20]

Erosion and Desertification

Soils left bare by clearing or logging are eroded by heavy rains and strong winds. Erosion is especially important where tropical rainforests are cut down because the thin layer of nutrient-rich soil can be lost quickly. Nutrient-poor soils can only support crops for a few years. Once the crops fail, farmers are forced to move to new areas and the cycle of deforestation begins again.

Natural forces such as wind, water, and erosion, combined with human activities such as agriculture, overgrazing, logging, and deforestation, can create desertlike conditions in areas that were once forested. This process is called desertification. When human populations were low, desertification took centuries or millennia. Now, desertification takes only decades or a few years. In the United States, rangelands are degraded by agriculture and overgrazing, causing dust storms and lower soil fertility. In some countries, desertification is accelerated when animal manure is used for fuel instead of left on the soil where it acts as a fertilizer.

Poor soils, heavy erosion, high species diversity, and short seed life prevent tropical rainforests from fully recovering. When small patches are cleared, bare soil is still shaded and close to seed sources. The forest in the small patch can regrow. But, when a large area is cleared, the same type of forest, with all its original species diversity, may never return. This is a key difference between clearing forests in the tropics and in temperate areas. With time, carefully replanted temperate forests can often regrow into forests with all the characteristics of the original forests. Apparently tropical rainforests cannot.

Facts About Tropical Deforestation

General

The Amazon River system is fed by more than 10,000 named tributaries.

Borneo receives up to 5 meters of rainfall each year.

Fifty to 80 percent of the moisture in the air above tropical forests comes from the forest itself through transpiration and evaporation.

Madagascar has 12,000 plant species. Sixty percent are found nowhere else on earth.

In 1990, tropical forests covered 1,756 million hectares. By region: South America and the Caribbean—52 percent (918 million hectares)

Africa—30 percent (528 million hectares)
Asia and the Pacific—18 percent (311 million hectares)[21]

One-half of the world's forests are in Brazil, Zaire, Indonesia, and Peru.

Deforestation

The average global deforestation rate for the tropics was 0.8 percent (15.4 million hectares) per year between 1980 and 1990.[22]

Annual deforestation rate by region:
   Asia and the Pacific—1.2 percent
   South America and the Caribbean—0.8 percent
   Africa—0.7 percent[23]

Tropical forest loss per year by region:
   South America and the Caribbean—7.4 million hectares
   Africa—4.1 million hectares
   Asia and the Pacific—3.9 million hectares[24]

On September 28, 1987, a satellite documented more than 5,000 separate fires in the Amazon.[25]

As of 1990, up to 240,000 fires were set annually during the dry season in the Brazilian Amazon.[26]

One-half of the world's original tropical forests have disappeared.

Brazil and Indonesia account for 45 percent of global rainforest loss.

The drug trade may destroy 33 percent of the remaining forest in Colombia by the year 2000.[27]

In 1991, in Kenya, there were 11,800 hectares of forest. In 1992, there were only 5,600 hectares. Forest covers only 3 percent of Kenya, but contains 50 percent of tree species, 40 percent of mammal species, and 25 percent of bird species.[28]

Myanmar's (Burma) deforestation rate jumped from 105,000 hectares in 1980 to 677,000 in 1990.[29]

All of Haiti's primary tropical rainforest has been destroyed.[30]

For every 6 acres that are deforested, an average of 1 acre is replanted as plantation.[31]

Partly because of international pressure, the amount of forested land cleared in Brazil dropped by 30 percent between 1988 and 1990.[32]

Japan receives more than 53 percent of the tropical hardwood trade. The United States receives 15 percent and Europe receives 32 percent.[33]

A single tropical rainstorm can wash away 185 metric tons of topsoil from each deforested hectare.[34]

Early European settlers deforested 1 million square kilometers in the United States. It caused the extinction of three bird species. The deforestation of a similar-sized area of Southeast Asia would likely cause the extinction of hundreds of birds, thousands of plants, and tens of thousands of insects.

Global Warming and Deforestation

The typical Amazon fire releases:
   4,500 metric tons of carbon dioxide
   750 metric tons of carbon monoxide
   25 metric tons of methane[35]

About one-third of the carbon dioxide released into the atmosphere from deforestation and burning each year is from the Amazon.[36]

Forest Uses

Indigenous people use tropical plants as anti-inflammatories, contraceptives, insect repellents, and muscle relaxants and to heal snakebites, wasp stings, intestinal worms, malaria, and fevers.

Three-fourths of all major drug plants used internationally were discovered through folk doctors and shamans.[37]

One study from the Peruvian Andes showed that forest uses, such as rubber tapping; extracting fruit, medicinal plants, and oils; and agroforestry provide seven times more income per hectare over a 50-year period than logging.[38]

Exporting countries would get five times as much money for wood if it were processed domestically rather than by exporting raw logs.

At least 1,400 tropical plant species are believed to contain anticancerous properties.

The Mexican yam contains a drug, diosgenin, used in birth control pills and steroids. In the mid 1970s, 180 tons of diosgenin was used each year for these purposes.

One-half of the wood extracted from the world's forests and woodlands is used for fuelwood.

One-half of the world's population uses wood for heating, cooking, and lighting. One hundred million people in the developing world can't get enough wood to meet their energy needs.

Worldwide, 1.3 billion people are using local fuelwood supplies faster than they can grow back.[39]

The katemfe fruit of west Africa has two proteins, each of which is 1,600 times sweeter than sucrose.

Tropical Deforestation Summary

Causes
some types of agriculture
some timber harvesting
fuelwood gathering
livestock grazing
road building
some government policies

Localized Causes
urbanization
mining
war
flooding caused by dam building

Consequences

extinction of some species (due to loss and fragmentation of habitat)

changes in climate and weather patterns (due to changes in the water cycle, changes in the albedo effect, and increased greenhouse gases stemming from deforestation)

erosion

desertification

Chapter 4
Biodiversity and Extinction

The Problem

Kinds of Biodiversity

Biodiversity is a combination of the words "biological" and "diversity." It means the variety of life and the ecological functions they perform (functions such as nutrient cycles, pollination, predation, photosynthesis, etc.). Biodiversity can be thought of in three ways: genetic diversity, species diversity, and ecosystem diversity. Genetic diversity is the variety of genes within a species. Varieties, subspecies, and breeds are expressions of genetic diversity. Species diversity is the variety of living organisms in the world. Ecosystem diversity is the variety of habitats and ecological processes in the natural world.

Species Diversity

Species are organisms that can reproduce among themselves, but cannot reproduce successfully with other organisms. Scientists estimate there are between 3 million and 30 million species on earth.[1] Of those, 1.4 million have been described (named) by scientists.[2] There are approximately 751,000 insect species, 9,000 birds, 4,000 mammals, 10,500 reptiles and amphibians, 248,000 higher plants, and a host of invertebrates and microorganisms known to exist.[3] However, recent studies have caused some scientists to increase their estimates. These researchers now believe there are as many as 30 million species of insects alone.[4]

Small species, such as insects, are discovered every day, but the discovery of new mammal species is newsworthy. In 1994, a new species of tree kangaroo was discovered in Indonesia, and a barking deer, named the giant muntjac, was found in Vietnam. Just two years earlier, a new species of wild cattle, the Va Quang ox, was discovered in the same area of Vietnam.[5]

Table 4.1—Number of Known Species (Major Groups)

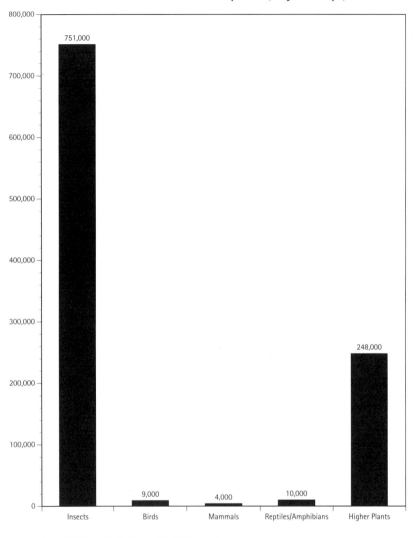

Source: E.O. Wilson, 1992, *The Diversity of Life*, W.W. Norton, New York.

Species variety is not the only aspect of species diversity that is important, however. A balanced mix of species is also key since a large number of non-native species is a sign of an unhealthy ecosystem.

Distribution of Biodiversity

Species diversity is unevenly distributed around the world. Tropical forests and coral reefs have a disproportionate number of species compared to temperate forests, deserts, and the Arctic. Between 40 and 90 percent of all species on earth live in tropical forests.[6] China, Indonesia, Brazil, Colombia, and Mexico are countries with a high diversity of mammal species. These countries, in addition to Ecuador and Venezuela, have a high diversity of flowering plants as well.[7]

Islands often have a large number of endemic species (species that are found nowhere else). Madagascar and Hawaii are good examples. Areas with Mediterranean climates also have high species endemism.

Scientists have identified eighteen areas with high concentrations of endemic species that are also threatened by habitat destruction. Of these eighteen "hot spots," fourteen are in the tropics and four are in Mediterranean climates.

The Benefits of Biodiversity

There are many convincing arguments for preserving species diversity, but probably the most persuasive argument is that people cannot survive without other living things. We depend on other organisms for food, medicine, industrial products, and a variety of ecological services.

Agricultural Products

Many foods are grown outside their country of origin. In the United States, 98 percent of crop production comes from species that originated elsewhere.[8]

People are dependent on wild plants and animals for new foods and genetic material. Plant breeding has been responsible for half of the increase in U.S. crop yields between 1930 and 1980.[9] Several crop varieties, grown together, can prevent total crop failure under harsh conditions. The foods we grow can be crossbred with wild varieties to develop more productive, drought-resistant, pest-resistant, and disease-resistant varieties. Wild plants with chemical defenses against herbivores can be used as natural pesticides in farming.

Medicines

Medicines are another reason biodiversity is important to people. One-fourth of the pharmaceuticals in the United States contain active ingredients derived from plants.[10] Many synthetic drugs were created by using natural chemicals as templates. Leukemia drugs, such as vinblastine and vincistine from the Madagascar periwinkle, are still extracted from the plant. An experimental drug from the venom of Malaysian pit vipers is being tested for use in breaking up blood clots and stopping strokes in progress. Eighty percent of people in developing countries depend on traditional medicinal plants and animals rather than modern medicines.[11] Pharmaceutical companies are beginning to tap into the knowledge of shamans and other traditional healers in their efforts to discover new drugs.

Industrial Products and Energy

Industrial products and energy are two other ways people are dependent on animals and plants. Fibers, natural rubber, waxes, and ingredients in mouthwash, detergents, explosives, candles, golf balls, and gum all come from the natural world. Forestry, fishing, and some tourism and recreation rely on animal and plant species. Trees, and in some cases manure, are important sources of energy around the world. Two billion people depend on fuelwood for cooking and heating.[12] Plants may become an even more important energy source as methanol and faster-growing fuelwoods are developed.

Ecosystem Services

Perhaps the most compelling argument for preserving living things is the ecosystem services they provide. The following services are fundamental to the normal functioning of the natural world.

Maintenance of the atmosphere—The balance of gases in the atmosphere is maintained in part by the nitrogen and carbon cycles. Plants and animals have a role in both of these cycles. In addition, forests influence global weather patterns (see Chapter 3).

Freshwater supplies—Plants increase the soil's capacity to absorb rainfall. Water is then released slowly into streams and rivers, thereby preventing erosion and flooding. Aquatic plants also filter out natural and human pollutants from water.

Soils—Plants and animals help to create soil by breaking down rock. Microorganisms help to maintain forest health by making nutrients available to plants.

Waste disposal—Decomposers dispose of organic wastes, recycle nutrients, and break down pollutants.

Pest and disease control—The populations of potential pests are often kept in check by other species. For example, when fur traders caused sea otter populations to decline in the Aleutian Islands, sea urchins became overpopulated. In turn, the abundance of sea urchins caused the decline of the kelp they fed on.[13]

Pollination—Bees and other insects pollinate many important food crops such as peas, peppers, strawberries, apples, peaches, and grapes.

Aesthetics and Ethics
Some arguments for preventing extinction are aesthetic and ethical reasons. People enjoy, perhaps even require, contact with animals and plants. We hike, bird-watch, scuba dive, and keep pets. Even city dwellers enjoy knowing that large expanses of untouched wilderness exist. Others believe that living things have the right to exist, and it is our responsibility to protect them from extinction.

Extinction
Extinction is a natural process. Of all the species that have ever lived on earth, 90 percent have become extinct.[14] The difference between species extinction in the past and today's extinction rate is that species are now disappearing much faster because of human activities. Over geologic time, the natural extinction rate, or the background rate, has been approximately one mammal species every 400 years and one bird species every 200 years.[15] The background rate is used to compare the average extinction rate with episodes of accelerated extinctions. When the background rate doubles, it is considered a mass extinction.

The extinction rate caused by humans in the last two centuries is unprecedented. The rate of extinction is now 25,000 times higher than the background rate.[16] Current projections are that 1 to 11 percent of species will be committed to extinction per decade between 1975 and 2015.[17] Many of these species have not and will not be described by science before disappearing. With them may pass an opportunity to discover a cure for AIDS or a disease-resistant food crop.

Another difference in today's extinction rate is that we are losing a larger proportion of terrestrial plants. Plant extinction is crucial since the extinction of one plant can lead to the loss of up to thirty animals and plants that depend on it.[18]

Some animal and plant species are more susceptible to extinction. Endangered ecosystems include ancient temperate rainforests, tallgrass prairies (in the United States), tropical rainforests, tropical coral reefs, coastal wetlands, and ancient lakes. Island species are more susceptible to human disturbance because they are confined to smaller areas and have evolved in isolation. This magnifies the impact of hunting, habitat destruction, fire, and the introduction of non-native species.

Human interference is accelerating the extinction rate of tropical rainforest species 1,000 to 10,000 times.[19] Tropical birds are especially sensitive to habitat disturbance because they have evolved in a relatively stable environment, have low reproductive rates, and are preyed upon heavily.

Species that live long lives but reproduce slowly (e.g., elephants), or are rare to begin with, are more likely to become extinct. In addition, carnivores are more vulnerable to extinction because they feed at the top of the food chain, are more susceptible to pesticide and heavy metal poisoning (see Chapter 5), and need to travel greater distances to find food.

Once the population of a species declines, it is more susceptible to disease, inbreeding, hunting, habitat destruction, and natural disasters. Many species are doomed to extinction because the population is no longer able to overcome these obstacles.

Human Causes of Extinction

There are direct and indirect causes of extinction. Direct causes include habitat loss, introduction of non-native species, overhunting, pollution, and some agricultural and forestry practices. Where the cause is known, 39 percent of extinctions since 1600 were caused by the introduction of non-native species, 36 percent by habitat destruction, and 23 percent by hunting and purposeful extermination.[20]

A major direct cause of species extinction today is habitat loss. As the world struggles to deal with population growth, political instability, poverty, and greed, habitats are damaged and destroyed. People pave over, flood, log, overgraze, plow, and pollute the natural areas that animals and plants need to survive. The result is complete loss, degradation, and fragmentation of habitat (dividing habitat into patches that are surrounded by cleared areas).

Agriculture is a widespread cause of habitat destruction. Forests are cut down and grasslands are plowed over to create agricultural

Figure 4.2—Percentage of Land Area Protected and Countries with No Protected Areas

	Percentage of land area protected	Countries with no protected areas
Ecuador	39.3%	Equatorial Guinea
Venezuela	30.2%	Guinea-Bissau
Austria	25.3%	Mauritius
Germany	24.6%	Cambodia
Dominican Republic	21.5%	Iraq
Butan	19.3%	Syrian Arab Republic
United Kingdom	18.9%	United Arab Emirates
Switzerland	18.2%	Yemen
Botswana	17.6%	Solomon Islands
Panama	17.2%	

Source: World Resources Institute, 1994, *World Resources 1994-95: A Guide to the Global Environment*, Oxford University Press, New York.

fields. People are creating deserts by diverting water, collecting fuelwood, and allowing land to be overgrazed.

Wetlands are also being lost as they are drained and filled for agriculture, housing developments, and waste dumps. Between one-fourth and one-half of the world's swamps and marshes have been destroyed. In the United States, Iowa has plowed more than 99 percent of its natural marshes.[21]

Roads, railways, canals, and buildings fragment habitats. For some species, these barriers confine them to smaller and smaller patches of land.

Introduced species also drive extinction. Introduced species, combined with habitat destruction, are particularly devastating on islands because native species cannot escape their competition.

The problems confronted on islands are often the same as those confronted in protected areas such as parks and reserves. In a sense, they are islands of natural habitat in a sea of farmland or urbanization. Parks in the United States have not provided full protection to species. Bryce Canyon National Park has lost 36 percent of its original species, and Mt. Rainier National Park has lost 32 percent of its original species. Larger parks like Grand Teton and Yellowstone have fared better; only 4 percent of the species have become extinct.[22]

Occasionally species become extinct through a combination of habitat loss and other factors, such as predator control, pest control, the fur industry, wildlife trade, and commercial hunting and fishing. In the past, unregulated commercial hunting and habitat loss caused the extinction of the passenger pigeon and almost caused the

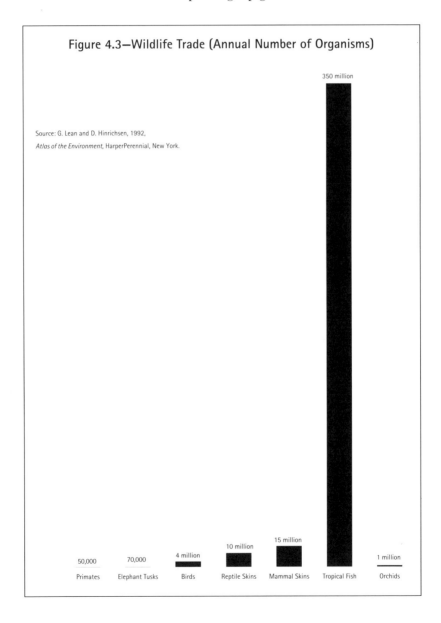

extinction of buffalo. Today, black rhinos and elephants are threatened by poaching for profit.

International trade of animals and plants accounts for up to $5 billion a year.[23] Major markets for illegally traded wildlife and wildlife products include the United States, Japan, and Europe. Worldwide, 622 species of plants and animals are threatened with extinction from trade. Illegal wildlife trade kills up to 80 percent of live animals before they even reach their final destination.[24]

A future threat to biodiversity is climate change. If precipitation patterns and temperatures shift, species that are not able to adapt or migrate will perish.

Habitat loss, introduced species, overhunting, pollution, and some agricultural and forestry practices are the direct causes of biodiversity loss. There are also deeper causes that drive biodiversity loss. These indirect causes, or root causes, need to be addressed if extinction is to be slowed.

One indirect cause of biodiversity loss is the combination of population growth and high resource consumption. Both the high population growth rate (primarily in developing countries) and high consumption patterns (primarily in developed countries) have put increasing pressure on the earth's natural resources and have aggravated habitat loss and pollution.

Another root cause of biodiversity loss is lack of knowledge about the natural world. Scientists have only discovered a fraction of the species that exist and there are huge gaps in the knowledge about how ecosystems function. Without that information, it is difficult to assess the full impact people have on biodiversity.

Economic systems and policies that do not take into account the value of biodiversity can be destructive. Ecosystem services provided by biodiversity, such as pollination, filtration of polluted water, and pest control, are not usually valued, while the products of biodiversity, such as timber products, medicines, and fuelwood, are. Typically the sustainable uses of biodiversity are not included in our economic systems. Government policies, such as tax incentives and property rights laws, also influence how people use land, animals, and plants.

Modern agriculture and trade patterns often discourage the maintenance of biodiversity. Today, many agricultural producers specialize in a few crops rather than an array of crops. Consumers demand a constant supply of a relatively few food varieties rather than a smaller supply of many unusual varieties.

Inequitable distribution of resources is another root cause of biodiversity loss. This is particularly dramatic on an international scale. Countries that are deep in debt look for ways to make money quickly. One way is to exploit their natural resources. But even on an individual scale, people who are desperate to survive do not have the luxury to consider the future when their immediate needs are so strong.

Consequences

When we consider all the benefits of biodiversity, it is not hard to imagine what the consequences of species extinction will be. Our food crops will become more difficult to maintain as wild genetic materials disappear. Already, up to 80 percent of the world's food supply is dependent on two dozen species.[25] If we lose their wild cousins, our opportunities for creating such crops as perennial corn and disease-resistant wheat are also lost. In addition, food crops will be affected as pollinator species go extinct.

The loss of biodiversity limits the possibility of a cure for cancer, AIDS, and other diseases. Industrial products and new energy sources disappear as wetlands and forests are plowed over.

The free ecosystem services provided by living things will disappear as species become extinct. The balance of atmospheric gases that keep our global climate and temperature in balance will be altered as forests are cut down. The interaction between predator and prey is a balance that prevents species from becoming pests. Plants that prevent erosion and filter out pollutants are being lost. We may also lose the services of microorganisms that dispose of wastes and keep soils fertile.

Perhaps the most dire consequence of extinction is the loss of the genetic material that species represent. Even when a population is saved from the brink of extinction, the population is more genetically uniform than it was previously. We may be forfeiting our choices when we drive a species to extinction. Along with the genetic material that is lost when a species becomes extinct, we may be losing a future food or a potential drug. It may take millions of years before the natural world recovers.

… Chapter 4: Biodiversity and Extinction

Facts About Biodiversity and Extinction
General
There are at most 1.4 million named species.[26] Scientific estimates of the actual number of living species range from 3 million to 30 million.[27]

If the information carried on one strand of DNA were converted into letters, it would fill all editions of the *Encyclopedia Britannica* published since 1768.[28]

Scientists predict that 40 to 90 percent of the world's total biodiversity exists in tropical forests. Countries with the greatest biodiversity include Brazil, Mexico, Madagascar, Indonesia, Colombia, China, Ecuador, Venezuela, and Zaire.[29]

A typical 10-square-kilometer patch of tropical rainforest contains up to 1,500 flowering plant species, 750 tree species, 400 bird species, 150 butterfly species, 100 reptile species, and 60 amphibian species.[30]

Each day, about twenty insect species are discovered by scientists.[31]

Eighty percent of Madagascar's plants are endemic. Madagascar has five times more tree species than temperate North America.[32]

There are 350 mammal species and 19,000 flowering plant species in the United States.[33]

California has more species than any state in the United States. Twenty-five percent of extinct and endangered species in the United States are in California.[34]

Only 10 percent of endangered and threatened species in the United States have increasing populations.[35]

Twenty-four plant and animal species account for 80 percent of the world's food supply.[36]

Ninety-eight percent of the crop species in the United States originated elsewhere.[37]

Ninety crops in the United States depend on insect pollinators.[38]

One-fourth of all drugs sold in the United States come from plant sources.[39]

A barley plant from Ethiopia contains a gene that protects a $160 million barley crop in California from lethal yellow dwarf virus.[40]

The Great Barrier Reef has at least 3,000 species.[41]

Two-thirds of fish caught worldwide are hatched in tidal zones.[42]

Forty-three ant species live on one leguminous tree in Peru. That is roughly equal to all the species of ants found in the British Isles.[43]

Large new species discovered since 1992 include:
Tree kangaroo—Indonesia 1994
Giant muntjac (barking deer)—Vietnam 1994
Va Quang ox—Vietnam 1994[44]

All the world's zoos can fit into an area the size of Washington, D.C.[45]

Extinction

Ninety percent of all species that have existed on earth have gone extinct. Most of them disappeared from natural causes.[46]

If current trends persist, 1 to 11 percent of the world's species will be committed to extinction each decade through the year 2015.[47]

A species is declared extinct 50 years after the last sighting.[48]

About 70 percent of mammals and birds that have become extinct in recent history lived on islands.[49]

The rate of extinction due to rainforest disappearance is one to 10,000 times faster than before human disturbance.[50]

Extinction in U.S. national parks:
Bryce Canyon—36 percent of the original species

Chapter 4: Biodiversity and Extinction

Mt. Rainier—32 percent of the original species
Yosemite—25 percent of the original species
Grand Teton and Yellowstone—4 percent of the original species[51]

The United States has 4,500 introduced species.[52]

Natural wetlands lost in the United States during the last 200 years:
Iowa—99 percent
California—91 percent
Ohio—90 percent[53]

Seventy-five percent of U.S. plant and bird extinctions have occurred on the Hawaiian islands.[54]

The world wildlife trade may be a $5 billion-a-year business. As much as one-third of it is illegal. In a typical year wildlife trade includes:
25,000 to 30,000 primates
2 to 5 million birds
1,000 to 2,000 tons of raw coral
7 to 8 million cactus plants
10 to 15 million raw reptile skins
500 to 600 million ornamental fish
9 to 10 million orchids[55]

From 60 to 80 percent of all live animals illegally traded die in transit.[56]

In 1990, white rhino horns sold for $12,500 a pound.[57]

Rare species of cactus can sell for $15,000 apiece.[58]

The population of black rhinos has decreased by 95 percent since 1970.[59]

The United States is the world's top importer of live primates, live parrots, and reptile skins.[60]

Biodiversity and Extinction Summary

Causes

habitat loss
introduction of non-native species
pollution
overhunting
some agricultural practices
some forestry practices
population growth
overconsumption of natural resources
some government policies

Consequences

extinction of species:
    loss of potential medicines
    loss of potential food crops
    loss of potential industrial products
    changes in ecosystem functions

Chapter 5
Pollution

The Problem
Historically, air, water, and land pollution has had local or regional effects. Now it is having global impacts. Tons of poisonous heavy metals (metals with specific chemical structures) flow to the North Sea from Germany's streams, every ocean of the world is littered, and clouds of factory smoke in one country form acid rain in another country. People in some areas of eastern Europe and the former Soviet Union have life expectancies that are years lower than in unpolluted areas; cancer rates are higher, and reproductive problems are more serious.

Types of pollution are not easily categorized. Air pollution affects soil, seawater, and freshwater. Litter and toxic waste can lie on the surface of the ground or water, and heavy metals can be found anywhere. This chapter separates pollution into three categories: land (primarily solid waste), water, and air.

Causes
Pollution on the Land
Solid Waste Disposal
In many countries, waste disposal means throwing trash in a nearby stream or on an empty lot. Waste disposal practices in industrialized nations are more regulated than in developing countries, but the amount of trash generated by each citizen is much higher. A resident of New York City, for example, throws away three to four times as much as a resident of Calcutta or Manila.[1]

Aluminum and plastic now make up a greater proportion of waste than more conventional solid wastes like glass, steel, and plant fibers. In the United States, plastics account for 9 percent of the total waste by weight and 20 percent by volume.[2] When buried, most plastics break down slowly and take up space. Worldwide, the popularity of plastic continues to grow. By the year 2001, the world could be consuming as much as 100 billion pounds of plastic each year.[3]

Industrial waste far exceeds any other category of waste generated in the United States. A full 93.7 percent of waste comes from industrial sources.[4]

Landfills, sanitary landfills, and incineration are some of the methods used to dispose of waste. While better than dumping trash on an empty lot, landfills and incineration also have drawbacks.

Landfills

Landfills are places where waste is buried in the ground. They are safer than open dumps, but industrial sludge and chemicals as well as household toxins can leak into the surrounding soil and water. One-half of the official landfills don't meet regulations in the former Soviet Union. In the United States, more than one-fifth of the waste sites on the Superfund cleanup list are municipal landfills. Landfills are also filling up and closing down. In 1988, 8,000 landfills were open in the United States. By 1991, only 5,812 landfills were open.[5]

Sanitary Landfills

These are landfills lined with heavy plastic. Each day a depression is filled with garbage, bulldozed over with fill dirt, and compacted down. The dangers of sanitary landfills include groundwater pollution, leachate (the chemical "soup" that forms in landfills) runoff, and methane production (a greenhouse gas and explosion hazard).

Incineration

Incineration means burning waste. If not done correctly, burning waste can release heavy metals, nitrogen, sulfur oxides (a component of acid rain), carbon monoxide, dioxins, particulates, and acid gases.

As more and more solid waste is produced, disposal will continue to be a problem for all countries. The amount of solid waste generated per person in the United States has risen from 2.7 pounds in 1960 to 4.3 pounds in 1991.[6] The U.S. Environmental Protection Agency (EPA) estimates that each year Americans throw away 242 million tires, 1.6 billion pens, 2 billion razors, and 16 billion diapers.

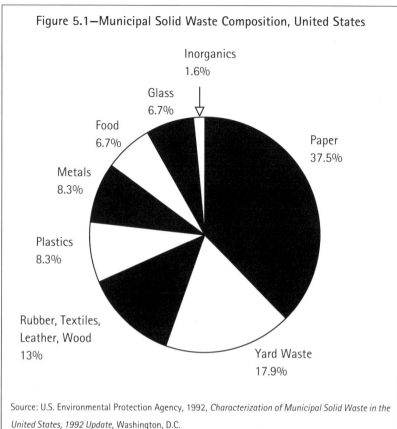

Figure 5.1—Municipal Solid Waste Composition, United States

Source: U.S. Environmental Protection Agency, 1992, *Characterization of Municipal Solid Waste in the United States, 1992 Update*, Washington, D.C.

Hazardous Waste

The EPA classifies hazardous wastes into four categories: ignitable, corrosive, toxic, and reactive (a substance that acts with another substance to create a chemical change). It lists 50,000 hazardous chemicals. Hazardous waste makes up 5 percent of the solid waste produced in the United States.[7]

Poland generates about 30 million metric tons of hazardous waste each year and two-thirds of that is dumped into unregulated waste sites.[8] Worldwide, hazardous wastes are routinely dumped into rivers, lakes, oceans, landfills, settling ponds, roadsides, and the air.

Sources of Hazardous Waste

Hazardous wastes come from industry, agriculture, and households. In the United States, about 70 percent of the toxic wastes come from the chemical industry.[9] Agricultural chemicals banned in the United States are still being used in other countries. They contaminate soils or accumulate in lake bottoms. Household toxins such as cleaners, solvents, paints, pesticides, motor oil, and batteries are buried or dumped onto the ground.

What Happens to Hazardous Wastes?

Once hazardous wastes are buried or dumped, they have several fates. Sometimes they leach through the soil as water, or snowmelt filters through the soil. This waste is carried deep into the soil or groundwater. Hazardous wastes can vaporize into the air from contaminated soil or move to new areas by wind or water.

People inhale and ingest contaminated soil particles when eating contaminated foods or eating with dirty hands. Children are particularly susceptible to ingesting contaminated dust while playing. Root vegetables accumulate poisonous heavy metals and some pesticides in their tissues, and windblown contaminated dust collects on fruits and vegetables. Grazing animals also ingest and inhale contaminated soil which meat-eaters ultimately consume.

Effects of Hazardous Waste

Over 70,000 chemicals are traded around the world and many of them have not been tested for their harmful effects.[10] Evidence of medical or environmental problems from any particular chemical is difficult to find.

People who drink water contaminated by high concentrations of hazardous waste can expect a variety of symptoms, including sore throat, dizziness, burning eyes, vomiting, diarrhea, rashes, headaches, cough, shortness of breath, nausea, weakness, and burning in the nose. Hazardous waste affects the environment by poisoning wildlife and water sources, and by inhibiting plant growth.

Pesticides

The agriculture industry uses 75 percent of the pesticides used in the United States. Government and industrial lands use 12 percent, as do households. The final 1 percent is used on forest lands.[11] Pesticides are used on crops for good reason. A full 42 percent of the U.S. food supply is lost to disease and pests each year.[12] But there are downsides to using pesticides. The U.S. National Research Council estimates that up to 20,000 Americans may die each year from cancer related to eating foods grown with pesticides.[13] As many as 50 million Americans drink pesticide-polluted water.[14] More than 250,000 Americans get sick from household use of pesticides each year.[15] But most pesticide problems occur in developing countries. The World Health Organization estimates that almost 3 million people, particularly agricultural workers in the developing world, suffer from acute pesticide poisoning each year. Dangerous pesticides banned in the United States show up in foods that are imported from countries with weaker restrictions.

Insect species targeted by particular pesticides often become immune. The pace of pesticide immunity is quickening and pesticide use is growing to keep up. The disheartening aspect of pesticide use is that the agricultural loss to pests (30 percent of the world's harvest) before and since the use of agrichemicals has remained the same.[16]

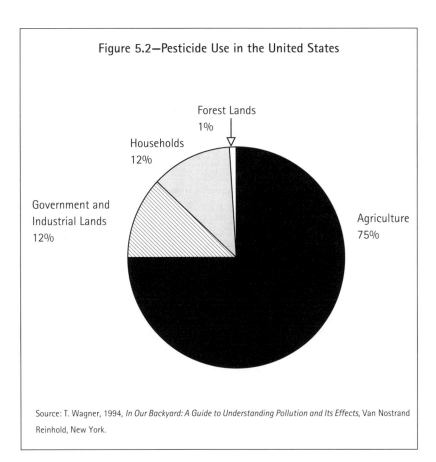

Figure 5.2—Pesticide Use in the United States

Forest Lands 1%
Households 12%
Government and Industrial Lands 12%
Agriculture 75%

Source: T. Wagner, 1994, *In Our Backyard: A Guide to Understanding Pollution and Its Effects*, Van Nostrand Reinhold, New York.

Exporting Waste

Even though international agreements regulate some worldwide waste transportation, shipments of waste to developing countries from industrial nations continue to increase. Wastes are shipped to save money and because landfill space is running out.[17] The receiving nations are paid to accept the waste. In 1989, an estimated 3 million tons of hazardous waste was being shipped around the world.[18] Once the waste reaches its destination, it is often dumped in unregulated waste dumps. Both the transport of waste across the oceans and developing country waste dumps threaten the health of people and the environment.

Air Pollution

Until the 1960s, air pollution caused problems locally and regionally. Now it is a global problem. Between 1980 and 1984, over 600 million people in the world lived in urban areas where sulfur dioxide concentrations exceeded the World Health Organization's recommended limits.[19] More than 1 billion people live with unhealthy levels of particulates.[20] In 1988, the air in Los Angeles exceeded health standards for 232 days and, in Mexico City, it exceeded health standards for 312 days.[21] Children brought up in the most polluted areas of Los Angeles are left with a 10 to 15 percent reduction of lung capacity for the rest of their lives.[22] Breathing Mexico City's air for one day is like smoking two packs of cigarettes.[23] And in Russia, government officials estimate that 20 to 30 percent of deaths are linked to environmental problems such as air and water pollution.[24]

Air pollution decreases visibility at its source and for hundreds of miles away. It can travel up to 300 miles a day. The distance we are able to see in the United States has declined by 25 percent from 1948 to 1983. Visibility has improved in the northeast United States, but has diminished by 80 percent in the Southeast. Air pollution from Los Angeles creates a haze over the Grand Canyon and other national parks in Utah, Nevada, and Arizona.[25]

The news on air pollution is not all bad, however. In the United States between 1985 and 1994, sulfur dioxide concentrations decreased by 25 percent, nitrogen oxides by 9 percent, carbon monoxide by 28 percent, ground-level ozone by 12 percent, particulates by 20 percent, and lead by 86 percent due in large part to the Clean Air Act.[26] The number of "poor air days" in many U.S. cities has dropped as well. Between 1984 and 1986, Denver had 147 "poor air days,"

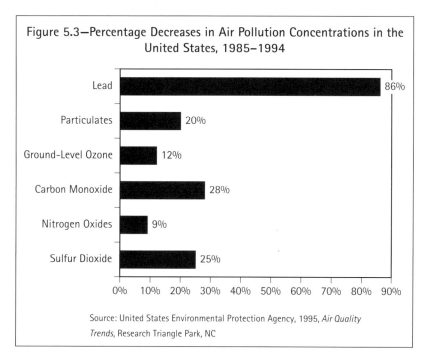

Figure 5.3—Percentage Decreases in Air Pollution Concentrations in the United States, 1985–1994

Lead 86%
Particulates 20%
Ground-Level Ozone 12%
Carbon Monoxide 28%
Nitrogen Oxides 9%
Sulfur Dioxide 25%

Source: United States Environmental Protection Agency, 1995, *Air Quality Trends*, Research Triangle Park, NC

New York had 210, and Sacramento had 195. Between 1991 and 1993, Denver had 17, New York had 34, and Sacramento had 45 "poor air days."[27]

The story is the same in some other countries. In Germany, air-pollution emissions have decreased since 1989. Carbon dioxide emissions are down by almost 50 percent, sulfur dioxide by 44 percent, and carbon monoxide by 32 percent.[28]

The Effects of Air Pollution

A person's sensitivity to air pollution depends on age, sex, health, level of activity, the outdoor temperature, and humidity. In general, older people, the very young, people in poor health, smokers, and those with chronic bronchitis, asthma, and heart disease are more susceptible. In addition to aggravating existing health problems, air pollution might cause lung cancer. Even minor air pollution can cause watery eyes, sore throats, and clogged sinuses.

Like humans, animals and plants need clean air to stay healthy. Damage to plants depends on the type and concentration of the pollutant, the species and maturity of the plant, light, temperature, moisture, nutrition, and the length of exposure. Forests and farms around polluted cities are showing the effects of air pollution. As early as 1942, crops close to Los Angeles were suffering. In 1960, $22,000 worth of lettuce was destroyed by smog in a single day.[29]

Air pollution damages human-made materials also. Paint, nylon, metal, and rubber erode. Clothing and buildings are soiled. Acid rain erodes marble and steel.

Particulates

Particulates are fine solids or liquids in the air. Smog is a combination of particulates and fog (as well as other pollutants, depending on the location). Smog forms when pollutants are not dispersed from a particular area. Calm winds, below-freezing temperatures, and cities where weather inversions form are factors that contribute to smog formation.

Sources

Particulates come from fires, volcanic eruptions, and incomplete combustion of fuels. Coal combustion is especially polluting. Liquid petroleum gas, used in Mexico as well as many other developing countries for cooking and heating, may be a major source of Mexico City's smog.[30]

Effects

Particulates and sulfur dioxide make a deadly combination, especially if they increase suddenly. The London smogs of 1952 and 1962 caused an increase in deaths from bronchitis and other diseases.[31] Human health is also harmed by long-term exposure to lower levels of particulates. The effects of particulates on human health depend on particulate size and whether they enter the lungs or the digestive system. They can cause lung damage, cancer, and premature death. Children, the elderly, and people with asthma, pulmonary, or cardiovascular disease are especially sensitive.

Particulates affect plants by covering leaf surfaces and blocking gas exchange. The particulates may also block sunlight, which inhibits plant growth.

Particulates reduce visibility and allow fogs to form more quickly. Buildings and fabrics are soiled and sometimes corroded when particulates accumulate on their surfaces.

Sulfur Dioxide (SO_2)

Sources

Power plants, industry, incinerators, and coal-burning fireplaces and furnaces are sources of sulfur dioxide. Lignite, a type of coal used in eastern Europe, is especially high in sulfur. Before reunification, East

Germany consumed more than 340 million metric tons of lignite each year. Forty-eight metric tons of sulfur dioxide was released into the air yearly as a result.[32]

Effects

When sulfur dioxide mixes with the air's moisture, acid rain and sulfuric acid mist form. Sulfuric acid causes chest pain, appetite loss, nosebleeds, coughs, sneezing, sore throat, shortness of breath, abnormal reflexes, and eye irritation.

Sulfur dioxide also causes severe damage to limestone, sandstone, iron, steel, copper, nickel, zinc, paper, leather, paint, medieval stained glass, fabric, and electrical wires. Greek and Italian monuments, which have sustained little damage for hundreds of years, are now corroding faster because of sulfur dioxide.

Sulfur dioxide is both helpful and harmful to plants. Sulfur dioxide provides needed sulfur to some, but other plants such as alfalfa, lettuce, spinach, peas, barley, and cotton are adversely affected.

Nitrogen Oxides (NO$_x$)

Sources

Nitrogen oxide is naturally produced by lightning, volcanic eruptions, and soil bacteria. Humans produce nitrogen oxide mostly through fossil-fuel combustion from motor vehicles, power plants, industry, and waste incineration.

Effects

One type of nitrogen oxide, nitrogen dioxide, causes a yellow-brown discoloration of the atmosphere. In people, nitrogen oxides cause burning eyes, headaches, and an increased susceptibility to respiratory problems.

Like sulfur dioxide, nitrogen oxide mixes with sunlight, oxygen, and water to form acid rain and acid fog. Nitrogen oxide is second to sulfur dioxide in its contribution to acid rain. In addition, nitrogen oxide reacts with hydrocarbons, oxygen, and sunlight to form ozone, which is damaging to plants and people. Finally, nitrogen oxide absorbs infrared radiation, which can contribute to global warming.

Carbon Monoxide (CO)

Sources

Carbon monoxide comes from the combustion of fossil fuels, industry, waste incineration, and biological processes. In polluted cities, carbon monoxide levels frequently correspond to traffic and weather conditions. Ten percent of carbon monoxide emissions come from burning wood in household fireplaces.[33]

Effects
Carbon monoxide is a colorless, odorless gas. It has no known effects on human-made materials or vegetation. In humans, however, carbon monoxide can cause a number of health problems. It is absorbed through the lungs where it reduces the ability of the blood to carry oxygen. High levels of carbon monoxide can impair vision; cause erratic behavior; aggravate heart problems; cause headaches, drowsiness, and fatigue; and may ultimately result in a coma and death.

Ozone, another pollutant, is created when carbon monoxide is exposed to sunlight.

Volatile Organic Compounds (VOCs)
Sources
Volatile organic compounds come from motor-vehicle exhaust, chemical manufacturers, solvents, paint shops, and dry cleaners. VOCs are carbon compounds that are reactive in the atmosphere.

Effects
VOCs react with nitrogen oxides and sunlight to form ground-level ozone. Ozone is toxic to people and is a greenhouse gas.

Toxic Chemicals
Sources
More than 70,000 synthetic chemicals are traded around the world.[34] Some of them are released into the atmosphere through waste disposal, sewage treatment, waste incinerators, industry, small commercial and residential sources, and agriculture. Worldwide, industry alone releases 2.7 billion pounds of toxic chemicals each year.[35]

Effects
The long-term effects of a single chemical are difficult to determine. Polychlorinated biphenyls, better known as PCBs, are one example of the problems with toxic chemicals. PCBs are used in carbonless reproducing paper, high-voltage transformers, epoxy paint, and coatings for wood, metal, and concrete. When materials containing PCBs are burned, the PCBs are released. They then accumulate in the tissues of animals and are passed through the food chain. Aquatic animals are particularly susceptible to PCB accumulation in their tissues.

Toxic "Heavy" Metals

Heavy metals are metals with particular chemical structures. Beryllium, lead, mercury, nickel, and cadmium are the major metals found in air pollution.

Sources

Toxic metals are released from motor vehicles, industry, and metal smelters. Lead also comes from large disturbances of lead-containing soil, leaded gasolines, and the manufacture of cable, solder, shot, paint, batteries, and lead pipe. Lead emissions have dropped dramatically since 1970 due to the decreased use of leaded gasoline and air pollution control devices. In 1970, 223,300 tons of lead was emitted. In 1991, only 5,467 tons was emitted.[36]

Effects

Other than lead, most toxic metals only have local effects. Lead is more widespread because it is used in fuel additives and industry. Automobile fuels in the United States contain less lead, but this is not the case in many other countries. Lead is poisonous to humans, plants, and animals. The symptoms of lead intoxication in humans are nausea, abdominal pain, anemia, vomiting, irritability, and clumsiness. Higher levels of lead damage the brain, kidneys, nervous system, and red blood cells.

Radioactivity

Sources

High-altitude flight, the phosphate industry, and using building materials containing radon, such as pumice and granite, increase exposure to radiation. In addition, people are exposed to radioactivity from medical sources, nuclear explosions, and nuclear power plants.

Effects

People are exposed to radiation externally and internally. Internal exposure comes from inhaling or ingesting radioactive materials. Radiation exposure causes abnormal cell development; damage to the cell's chromosomes; leukemia; cancer of the thyroid, lungs, and breasts; slowing of growth and development; and death. Since the Chernobyl disaster, death rates in northern Ukraine have increased 16 percent and the Russian Chernobyl League estimates that 7,000 cleanup workers have died or committed suicide as a result of the Chernobyl disaster.[37]

Photochemical Pollution

Sources

When carbon monoxides, nitrogen oxides, and hydrocarbons are exposed to sunlight, new chemicals form. These secondary pollutants make up photochemical smog. This reaction with sunlight produces ozone, peroxyacetyl nitrate (PAN), and other chemicals.

Effects

Photochemical smog causes eye irritation, coughs, chest discomfort, sore throats, stuffy noses, and headaches. Ozone, in particular, is more toxic during exercise because more of it is inhaled. Ozone also causes severe fatigue and affects visual acuity. Crops such as spinach, tomatoes, pinto beans, tobacco, and ponderosa pine trees are particularly sensitive to ozone. Damage to California crops from ozone is estimated at $1 billion a year.

Ozone is partly responsible for a decrease in visibility and acts as a greenhouse gas when it is in the lower atmosphere. Unfortunately, ozone in the lower atmosphere does not travel to the stratosphere where it could replenish the depleted ozone layer. Ozone molecules are not stable enough to travel to the upper atmosphere.

Acid Rain

Sources

Acid rain is created when sulfur oxides, nitrogen oxides, and hydrogen chloride (from burning fossil fuels) combine with water vapor. The result is a damaging mixture of sulfuric acid, nitric acid, and hydrochloric acid. This mixture is then flushed out of the atmosphere by rain, snow, or in some cases, dry particles. Volcanic eruptions, ammonia, soil particles, and sea spray also contribute to acid rain, but industrial pollution is the major contributor. In some areas, pollution increases the acidity of rain 100 times the natural level. Mountains are particularly susceptible to acid rain because they encourage waterlogged clouds to release rain and snow. Before the 1960s, most acid rain fell near emission sources such as the Ohio River Valley. Taller stacks on factories now send emissions to locations farther away.

Measuring Acidity

Acidity and alkalinity are measured using the pH scale. The pH scale runs from 0 to 14; 7 is neutral. The more acid a substance is, the lower the number. The more alkaline, the higher the number. Every point on the scale increases by a factor of ten. For example, a pH of six is ten times more acidic than a substance with a pH of seven and a pH of five is 100 times more acidic than a substance with a pH of seven. Normal rainwater usually has a pH higher than 5.6. Healthy lakes and rivers have a pH ranging from 5.6 to 8. Some polluted areas of Japan, central Europe, the northeast United States, and Scandinavia have rainfall with a pH as low as 3.5.[38]

Effects
The effects of acid rain depend on the acidity of the rain and the neutralizing power of the lake, soil, or material that is being exposed to precipitation. Acid rain and acid fog corrode buildings made of marble, stained glass, and metals.

Acid rain acidifies river and lake water, killing many living organisms. Ten percent of the lakes in the Adirondacks are too acidic for fish.[39]

Acid rain also changes soil fertility by disrupting the balance of organisms, nutrients, and toxic metals in soils. Plant nutrients such as potassium, magnesium, and calcium are leached from the soil by acid rain. Alkaline soils like those of the midwest United States and southeast England are able to neutralize acid rain. Soils high in calcium and lakes with limestone or sandstone beds neutralize acidity.

Acid rain may be partly responsible for forest die-offs around the world. Called "Waldsterben" in Germany, the symptoms of forest death include yellow needles, needles that fall off, bark damage, damaged treetops, and stunted growth. Eighty-two percent of Poland's forests, 78 percent of Bulgaria's forests, and 73 percent of Czechoslovakia's forests were damaged by 1989.[40] Pine forests stretching from Georgia to New England are also dying. On North Carolina's Mt. Mitchell, two-thirds of the living trees have lost leaves and needles.[41] However, acid rain alone may not be killing the forests. Ozone, carbon monoxide, and natural stresses could be acting together. Insects, frost, and wind deliver the final blow.

Indoor Air Pollution
Sources
Tobacco smoke, asbestos (from insulation, roofing, and fireproofing), cleaners, combustion (from household appliances), radon (from natural radioactive gas seeping into buildings from rock and soil), and formaldehyde (from glues, plywood, insulation, and particle board) are all sources.

Effects

Indoor air pollution has become more problematic because the average person now spends 90 percent of their time indoors and because many buildings are weatherized, trapping pollutants inside.[42] Indoor air pollutants cause irritation of the eyes, nose, and throat; wheezing; coughing; fatigue; and skin rashes. Sometimes people become more sensitive to indoor air pollutants over time. Indoor air pollution may be partly to blame for the 40 percent rise in the incidence of asthma between 1982 and 1992.[43] Radon may cause 10 percent of lung cancer deaths in nonsmokers.[44]

Freshwater and Ocean Pollution

Freshwater is a necessary but relatively scarce resource. Less than 1 percent of the earth's freshwater is easily accessible to humans from lakes, streams, groundwater, and the earth's atmosphere, and that 1 percent is threatened by a barrage of pollutants. For example, groundwater is polluted by septic tanks, agriculture, landfills, hazardous waste sites, road salting, mining, chemical spills, and animal feedlots. In 1991, the U.S. Environmental Protection Agency (EPA) estimated that 1 million underground storage tanks are leaking.[45] In Poland, 60 percent of the river water is too corrosive to use for industry and one-half of its cities do not treat drinking water.[46] In other areas, water quality is improving. For example, organic compounds in Baltic Sea tributaries have decreased by 46 percent.[47]

Pollution of drinking water is particularly important. Diarrhea caused by unsafe drinking water kills over 4 million children a year.[48] Viral diseases, such as hepatitis, polio, gastroenteritis, and various diarrheal diseases, are passed through human populations by poor drinking water. Contaminated drinking water is also responsible for an assortment of other health problems, including sore throats, headaches, dizziness, cough, nausea, loss of taste, muscular weakness, burning eyes and nose, ringing in the ears, and shortness of breath. Coliform bacteria (types of bacteria carried in the feces of people and other warm-blooded animals) are monitored in water supplies to indicate water quality.

Deep oceans are still relatively clean, but litter and oil are found oceanwide. Seas that are enclosed (such as bays and sounds) or seas with slow currents are particularly vulnerable to pollution.

The sources of water pollution are often organized into two general categories: point sources and nonpoint sources. Point sources are sources where pollution discharge is measurable. They include industrial plants and sewage pipes. Nonpoint sources are sources where pollution output is difficult to measure, like agricultural run-off, storm water from cities, and chemicals from road deicing.

Dumping
Open bodies of water, especially the oceans, are often used as convenient dumping sites for waste.

Sources
Many substances are dumped into water, including toxic chemicals, sludge from water treatment plants, litter, war materials like grenades and mustard gas, radioactive waste, raw sewage, medical waste, and dredged material from harbors.

Effects
Waste dumped into the ocean decomposes, sinks to the bottom, or floats on the surface until it washes ashore. Waste that comes in contact with people and other living organisms poses a health risk.

Plastics are particularly dangerous to marine animals. Millions of seabirds, whales, seals, and dolphins are killed every year when they ingest or become entangled in plastic nets and containers.

Oil
Every year, about 4 million tons of oil pollute the oceans. More than one-fourth of it is intentionally dumped by ships, although deliberate dumping decreased in the 1980s.[49]

The largest accidental oil spill was the Ixtoc spill off the coast of Mexico in 1979. It leaked 400,000 metric tons, the weight of 73,333 elephants, over a span of 10 months.[50] But the cumulative effect of the small daily oil spills is just as damaging.

Sources
As much oil leaks into the ocean from natural undersea oil seeps as from oil-spill accidents; however, accidents are concentrated, sudden, and very damaging. Some oil reaches the ocean by air when fossil fuels are not completely burned. Oil also comes from domestic wastes and city streets. Finally, ships, refineries, petrochemical plants, offshore oil drilling, and dumping of industrial and automobile oils contribute to water pollution.

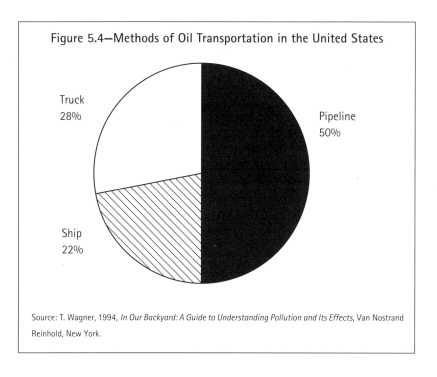

Figure 5.4—Methods of Oil Transportation in the United States

Truck 28%
Pipeline 50%
Ship 22%

Source: T. Wagner, 1994, *In Our Backyard: A Guide to Understanding Pollution and Its Effects*, Van Nostrand Reinhold, New York.

Effects

The fate of ocean oil spills depends on the type of oil and ocean conditions. In general, seven things can happen to oil once it is spilled:

1) Sinking—Barnacles and other organisms attach to tar balls, making them heavy enough to sink.
2) Breaking apart—Wind churns the seawater, breaking the oil slick into smaller clumps.
3) Evaporation—Evaporation takes months, perhaps years. The potential for evaporation is best during the hours immediately following a spill. Warm temperatures and waves help. Gasoline can completely disappear through evaporation, but heavy heating oil is not easily evaporated. The oil that does evaporate pollutes the air and may return to the ocean during rain showers.
4) Spreading—Oil is spread over the ocean's surface by wind and water currents. Light oil spreads more quickly than heavy oil.
5) Biodegradation—Some oil is broken apart by living organisms. There are many kinds of bacteria and fungi that partially decompose oil.

6) Washing ashore—Wind can push an oil spill onto the shore where the oil penetrates the sand or sediments. If the oil becomes buried where there is no oxygen, most of it will not break down.
7) Clumping—Some components in oils are not broken down. This oil often becomes balls of tar.

Of the 10.6 million gallons of oil spilled in the Exxon *Valdez* oil spill in 1989, 50 percent biodegraded, 20 percent evaporated, 14 percent was recovered, 12 percent sunk to the ocean bottom, 3 percent wound up on the shoreline, and 1 percent remains in the water.[51]

Oil spills affect marine animals in several ways. Natural ocean chemicals, used by some animals to navigate, are overpowered by oil. Fish eggs, floating on the ocean surface, can be killed by oil. Oil is especially damaging to marine birds. Oil-soaked birds cannot fly,

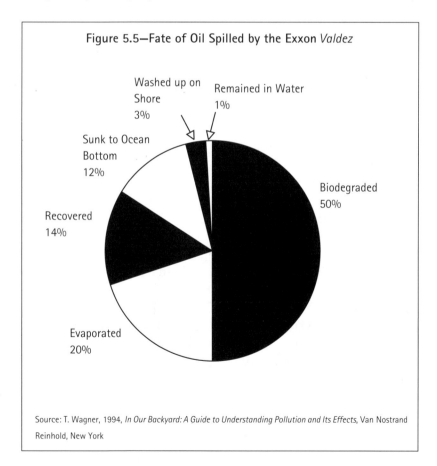

Figure 5.5—Fate of Oil Spilled by the Exxon *Valdez*

Source: T. Wagner, 1994, *In Our Backyard: A Guide to Understanding Pollution and Its Effects*, Van Nostrand Reinhold, New York

insulate themselves, or stay afloat. Furthermore, oil is ingested during preening and also seeps into eggshells.

Air Pollution Fallout
Fallout from air pollution is the most widespread type of water pollution. Approximately one-third of ocean pollutants fall from the air.[52] Air pollution fallout originates from dust, incomplete burning of fossil fuels, industry, agriculture, landfills, and radioactive fallout.

Thermal Pollution
Sources
Thermal pollution comes from industrial and electrical plant discharges, removal of shade trees along stream banks, and light-absorbing sediment particles that cause water to warm.

Effects
A body of water is thermally polluted when it receives so much heat that animals and plants are not able to adjust. Thermal pollution decreases the concentration of oxygen available for aquatic organisms, but warmer water also increases the possibility of algal blooms. Heat accelerates the metabolism of aquatic animals and lowers their resistance to disease. Some aquatic animals delay traveling to appropriate wintering grounds because of warmer water temperatures, leaving them stranded if the warm water stops flowing.

Radioactive Pollution
Sources
Radioactivity comes from natural sources, but it is also released into water from nuclear plants and during nuclear testing. For instance, from 1944 to 1971, the Hanford Nuclear Power Plant in Washington used Columbia River water as a coolant. The dozens of radioactive substances released each month flowed to the Pacific Ocean 360 miles away.[53] From 1946 to 1970, 86,758 containers of low-level radioactive waste were dumped off the coasts of the United States.[54] Britain dumped nearly 75,000 tons of radioactive waste into the North Atlantic between 1949 and 1982.[55] Airborne particles of radioactive plutonium released during nuclear tests are washed out of the air when it rains.

Effects
Radioactivity in water has similar effects to radioactivity in the soil or air: abnormal cell development; leukemia; cancers of the lung, breast, and thyroid; slowing of growth and development; and death.

Heavy Metals
Heavy metals are metals with a specific chemical structure. Heavy metals such as lead, mercury, cadmium, nickel, arsenic, zinc, copper and others are released into water by natural processes, but humans are also discharging heavy metals into rivers, lakes, and oceans. The Elbe River of Germany empties into the North Sea, carrying with it 10 tons of mercury, 24 tons of cadmium, and 142 tons of lead each year.[56] In addition to radioactive waste, Britain dumped at least 1 million tons of munitions, chemical weapons, and nerve gas at sea between 1949 and 1982.[57]

Sources
Heavy metals come from industry, car exhausts, mining, and smelters. The largest sources of lead are fossil fuels with lead additives, waste, and large disturbances of lead-containing soil. Mercury, lead, PCBs, dioxin, and cadmium enter the Great Lakes from power plants, garbage incineration, cars, insecticides, refineries, and steel mills.[58]

Effects
Heavy metals cause health problems in humans, other animals, and plants. The symptoms of lead intoxication in people are nausea, anemia, abdominal pain, clumsiness, vomiting, and loss of intelligence (especially in children). In more severe cases it can damage kidneys, brains, and red blood cells. Mercury causes numbness of the lips, arms, and legs; an irregular walk; narrowing of vision; and a change in touch, hearing, and speech. Mercury accumulates in the tissues of living organisms and is passed through the food chain. It affects the central nervous system and kidneys of people.

Agricultural and Urban Runoff
Sources
Fertilizers, pesticides, animal waste, and soil wash off agricultural lands. Chemicals and salt from road deicing, oil from automobiles, and waste from industry and domestic sources are washed off city streets by rain and snow and end up in streams, rivers, and lakes. Agricultural runoff is the greatest source of surface water pollution.[59]

Farmers in the United States use 24.2 billion pounds of chemical fertilizers and 1.8 billion pounds of chemical pesticides.[60]

Effects

Nitrogen and phosphorus from fertilizers, sewage, and industry cause algae blooms in surface water, and agricultural chemicals seep through the soil into freshwater aquifers. Some of these chemicals cause tumors and birth defects in aquatic organisms. Soil erosion from farmland and urban construction adds sediments to lakes and rivers. Sediments reduce water visibility and increase water temperature by absorbing more light. Urban areas pollute water when waste is funneled directly into water and when precipitation washes oils and chemicals off roadways.

Eutrophication

Eutrophication is a natural process that is accelerated by nutrient-rich runoff. It begins with an overabundance of nutrients in water. These nutrients feed algae, which multiply until the algae bloom blocks out sunlight needed by other plants. Some of the algae begin to die and decay, which decreases the amount of oxygen in the water. The decrease in sunlight and oxygen eventually kills more algae and many other organisms. At the end of the eutrophication process, the water is unable to support most living organisms.

Although eutrophication and algae blooms are becoming more common in lakes, it is unlikely that eutrophication will affect entire oceans. Nevertheless, oceans are suffering from more algae blooms. By 1992, Japan's Seto Inland Sea was experiencing 200 toxic algae blooms a year.[61]

Sources

Nutrients come from sewage treatment plants, agriculture, septic tanks, and urban runoff. These nutrients can cause eutrophication.

Effects

Eutrophication eventually kills many living organisms in a body of water. In less extreme cases, only highly sensitive organisms die from low oxygen levels and toxic algae.

Facts About Pollution

Pollution on the Land
Solid Waste

In the United States, plastics make up 20 percent of solid waste by volume and 9 percent by weight.[62]

Residents of New York City throw away four times more waste than residents of Calcutta or Manila.[63]

In 1960, Americans generated an average of 2.7 pounds of municipal solid waste per person per day. In 1991, each person generated an average of 4.3 pounds per day.[64]

Half of Japan's waste is recycled, one-third is incinerated, and less than one-fifth is buried.[65]

In 1988 there were 1,050 curbside recycling programs nationwide. By 1991 there were 3,955. During the same time, landfill space decreased 27 percent.[66]

Hazardous Wastes

The United States creates 240 metric tons of hazardous and toxic wastes each year.[67]

More than 20 percent of hazardous waste sites on the U.S. Superfund cleanup list are municipal landfills.[68]

Air Pollution

Air pollution can travel up to 300 miles a day.[69]

The daily fluctuations of air pollution in the deserts of Nevada and Arizona correspond to the pollution emitted during the Los Angeles workweek.[70]

Children raised in the most polluted areas of Los Angeles have a 10 to 15 percent decrease in lung capacity for life.[71]

Distribution of energy use in the United States:
energy production (electricity)—37 percent
transportation—27 percent
industry—24 percent
residential—12 percent[72]

The number of vehicles on the road in the United States increased 128 percent between 1969 and 1990.[73]

In 1988, smog levels in Mexico City exceeded World Health Organization standards for 312 days.[74]

More than 600 million people around the world live in urban areas where sulfur dioxide exceeds the World Health Organization's recommended standards.[75]

Over 1 billion people worldwide live in areas with unhealthy levels of particulates.[76]

The life expectancy in most polluted areas of Czechoslovakia is 5 years less than the national average.[77]

Plants susceptible to damage from sulfur dioxide pollution include alfalfa, lettuce, rhubarb, spinach, barley, cotton, and sweet pea.

Lead emitted from leaded gasoline, industry, and solid waste combustion dropped from 223,300 tons in 1970 to 5,467 tons in 1991.[78]

Percentage decrease in U.S. air pollution concentrations between 1985 and 1994:
sulfur dioxide—25 percent
nitrogen oxides—9 percent
carbon monoxide—28 percent
ground-level ozone—12 percent
particulates—20 percent
lead—86 percent[79]

Number of "poor air days":

	1984–1986	1991–1993
Seattle	38	0
Minneapolis	55	1
Denver	147	17
New York	210	34
Sacramento	195	45[80]

Declines in pollution levels in eastern Germany since reunification (1989):
sulfur dioxide—44 percent
carbon monoxide—32 percent[81]

Forest Death and Acid Rain

Half of the mid- and high-elevation red spruce trees of Vermont died from acid rain between 1964 and 1979.[82]

All bodies of freshwater in Sweden are unnaturally acidic. In 1984, Sweden attempted to neutralize acid lakes with lime at a cost of $25 million.[83]

By 1983, 8 percent of the former West Germany's forests were damaged by air pollution. By 1988, 50 percent of the forests were damaged.[84]

Damage to Europe's forests by 1989:
Poland—82 percent damaged
Bulgaria—78 percent damaged
Czechoslovakia—73 percent damaged
East Germany—57 percent damaged
Hungary—36 percent damaged[85]

Freshwater and Ocean Pollution

Less than 1 percent of the earth's freshwater is easily accessible to people from lakes, streams, groundwater, and water vapor in the atmosphere.

Only 40 percent of Czechoslovakia's waste water is sufficiently treated.[86]

Sixty percent of Poland's river water is too corrosive for industrial use.[87]

Ohio's Cuyahoga River was so polluted in 1969 that it caught fire.

Half of the people in the developing world don't have safe drinking water. Three-fourths have no sanitation facilities.[88]

About 70 percent of India's surface water is polluted.[89]

The Environmental Protection Agency estimates that one in five underground storage tanks in the United States is leaking. More than half of the U.S. population depends on groundwater for drinking.[90]

Method of oil transportation in the United States:
  50 percent transported by pipeline
  28 percent transported by trucks
  22 percent transported by ship[91]

Fate of the 10.6 million gallons of oil spilled by the Exxon *Valdez* oil spill in 1989:
  50 percent biodegraded
  20 percent evaporated
  14 percent was recovered
  12 percent sunk to the ocean bottom
  3 percent washed up on the shorelines
  1 percent stayed in the water[92]

Sources of ocean pollution:
  44 percent released directly into the sea or through rivers
  33 percent from air pollution fallout
  12 percent from shipping
  10 percent from ocean dumping
  1 percent from offshore drilling[93]

Every year, about 4 million metric tons of oil pollute the sea.[94]

Pesticide use in the United States:
  agriculture—75 percent
  government and industry land—12 percent
  household—12 percent
  forest land—1 percent[95]

Each month during 1968, 14,000 tons of waste water from pesticide plants along the Mississippi River washed into the Gulf of Mexico.[96]

Each year U.S. farmers use:
  24.2 billion pounds of chemical fertilizers
  1.8 billion pounds of chemical pesticides

In addition, 600 million pounds of chemical pesticides are used in households.[97]

In 1975, an estimated 6 million tons of solid waste was dumped into the ocean from ships.[98]

From 1944 to 1971, sixty types of radioactive materials traveled from Washington's Hanford reactor to the Pacific Ocean by way of the Columbia River.[99]

From 1946 to 1970, 86,758 containers of low-level radioactive waste were dumped off U.S. coasts.[100]

As of 1990, the Elbe River in Germany annually carried to the North Sea:
10 tons of mercury
24 tons of cadmium
142 tons of lead[101]

Since reunification, organic compounds flowing in Baltic Sea tributaries have dropped by 46 percent.[102]

Land Pollution Summary

Causes
solid waste
hazardous waste
pesticides

Consequences
human health effects of hazardous waste and pesticides (sore throats, rashes, headaches, coughs, shortness of breath, nausea, weakness, burning in nose)
environmental effects (poisoning wildlife, water pollution, can inhibit plant growth)

Air Pollution Summary

Particulates: Consequences
human health (lung damage, cancer, premature death)
environmental effects (can inhibit gas exchange on plant surfaces, corrodes buildings and fabrics)

Sulfur Dioxide: Consequences
human health (chest pain, appetite loss, nosebleeds, coughing, sneezing, sore throat, shortness of breath, abnormal reflexes, eye irritation)
environmental effects (corrodes buildings, some plants adversely affected while others benefit from additional sulfur)

Nitrogen Oxide: Consequences
human health (burning eyes, headaches, susceptibility to respiratory problems)
environmental effects (contributes to acid rain, forms ozone, can contribute to climate change)

Carbon Monoxide: Consequences
human health (impairs vision, erratic behavior, aggravates heart problems, headaches, drowsiness, fatigue, coma, death)

Volatile Organic Compounds (VOCs): Consequences
environmental effects (form ground-level ozone)

Toxic Chemicals: Consequences
environmental effects (accumulate in the tissues of animals, passed through the food chain)

Toxic "Heavy" Metals: Consequences
human health effects of lead (nausea; abdominal pain; anemia; vomiting; irritability; clumsiness; loss of intelligence; damage to brain, kidneys, nervous system, and red blood cells)
environmental effects of lead (toxic to plants and animals)

Radioactivity: Consequences
human health (abnormal cell development; leukemia; slowing of growth and development; death; cancer of thyroid, lungs, and breasts)

Photochemical Pollution: Consequences
human health (eye irritation, coughs, chest discomfort, sore throats, stuffy noses, headaches, severe fatigue, changes in vision)
environmental effects (decreases visibility, some crops affected, acts as a greenhouse gas in the lower atmosphere)

Acid Rain: Consequences
environmental effects (corrodes buildings, acidifies rivers and lakes, changes soil fertility, damages plants)

Indoor Air Pollution: Consequences
human health (eye, nose, and throat irritation; wheezing; coughing; fatigue; skin rashes; allergies)

Water Pollution Summary

Dumping: Consequences
depends on the type of waste (plastics can entangle marine animals, some wastes pose a health risk to humans and other animals)

Oil: Consequences
environmental effects (coats marine animals and birds)

Air Pollution Fallout: Consequences
depends on the type of waste

Thermal Pollution: Consequences
environmental effects (decreases concentration of oxygen in water, algae blooms, lowers disease resistance in some aquatic animals)

Radioactive Pollution: Consequences
human health (abnormal cell development; leukemia; slowing of growth and development; death; cancers of thyroid, lungs, and breasts)

"Heavy" Metals: Consequences
human health effects of lead (nausea; anemia; abdominal pain; clumsiness; vomiting; loss of intelligence; damage to kidneys, nervous system, brain, and red blood cells)
human health effects of mercury (numbness of lips, arms, and legs; irregular walk; narrowing of vision; changes in touch, hearing, and speech)

Agricultural and Urban Runoff: Consequences
environmental effects (algae blooms, tumors and birth defects in aquatic organisms, sedimentation of rivers and lakes)

Chapter 6
Global Population Growth

The Problem

The common denominator for ozone depletion, climate change, deforestation, biodiversity loss, and pollution is global population growth and the impact that each of us has on the environment. Every new birth means more natural resources are needed and the environment is stressed a little more. We may have already exceeded the earth's capacity to support the human population without causing irreversible damage to the environment.

The human population is growing exponentially. It took thousands of years for the human population to reach 1 billion people. It took 100 years to reach 2 billion people and it took only 30 years to add another billion. To increase from 3 billion to 4 billion took only 12 years.[1] In mid-1995, the global population was 5.7 billion. The growth rate is 1.7 percent. We are increasing by three people per

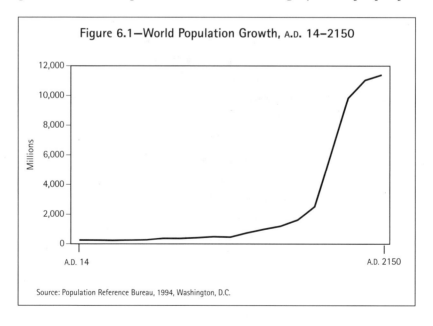

Figure 6.1—World Population Growth, A.D. 14–2150

Source: Population Reference Bureau, 1994, Washington, D.C.

second, or a quarter million every day.[2] The United Nations projects the world population will be 8.5 billion by 2025 and 10 billion by 2050.[3] Even deaths from AIDS will barely impact population growth. AIDS deaths for the entire decade of the 1990s are only expected to match 1 month of global population growth.[4] Population is expected to level off eventually. Current projections are that the global population will stabilize at 8–12 billion in the next century. The difference between the high and low projections depends on actions taken today.

Urban areas are especially impacted. In 1880, only 50 million people (or 5 percent) lived in urban areas. By 1985, 2 billion people lived in urban areas. Currently, 45 percent of the world's population lives in cities.[5]

From 1750 to 1930, the world's population grew faster in what is now the developed world. Since the 1930s, however, population growth in developing countries has outpaced growth in the developed world.[6] Now, 90 percent of the future population growth is expected to be in the developing world.[7] Austria, Belgium, and Italy have reached zero population growth. In Germany and Hungary, the population is actually shrinking.

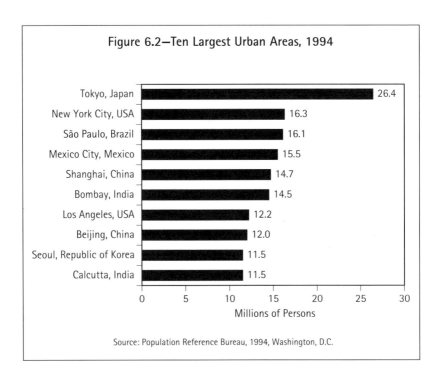

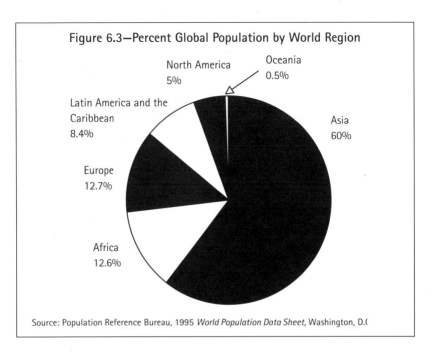

Figure 6.3—Percent Global Population by World Region

Source: Population Reference Bureau, 1995 *World Population Data Sheet*, Washington, D.C.

The United States is a special case among developed countries. The average number of children per woman in the United States is 2.1—a number that usually represents zero population growth.[8] However, the United States is the fastest-growing industrialized nation primarily because of immigration. The U.S. population is growing by 2.7 million people per year, which is equivalent to adding the population of Philadelphia or Washington, D.C., each year.[9] In 1994, the population of the United States was 260 million. By the year 2000, the population is projected to reach 275 million, more than twice the 1940 population. By 2050, the population is projected to increase by another 133 million, which is like adding thirty-eight cities the size of Los Angeles.[10] Some of the consequences of that growth may already be recognizable. The costs of education, water, highways, health care, and waste disposal are becoming increasingly difficult to withstand.

Causes

The main reason the world's population is getting bigger is that the death rate has decreased while the average worldwide birth rate has changed more slowly.

In the mid-19th century, the life expectancy was the same in the Northern and Southern Hemispheres. Now the average for countries in the North is 76 years, but only 60 years in the South. Even with this discrepancy between North and South, life expectancies have increased dramatically around the world. In the early 1950s, the life expectancy in China was 40.8 years, in India it was 38.7 years, in Indonesia it was 37.7 years, while the average life expectancy in developed countries was 66 years. By the late 1980s, the life expectancy in China rose to 69.4 years, in India it increased to 57.9 years, in Indonesia it jumped to 60.2 years, and in developed countries it rose to 74 years.[11]

Life expectancies have increased because food production has expanded, starvation and malnutrition have declined, sanitation and personal hygiene have improved, and modern medicine has helped people live longer, healthier lives. Major diseases, such as smallpox, cholera, plague, malaria, and tetanus, have been controlled in many areas. Infant mortality has also declined, but infant mortality is still more prevalent in developing countries.

In general, the birth rate in developing countries has stayed high while the birth rate in industrialized nations has steadily decreased. In many developing countries, couples are often motivated to have more children because they realize that some of their children will die. In rural areas, more children mean more food can be grown and more work gets done. Although the family may not rise above poverty,

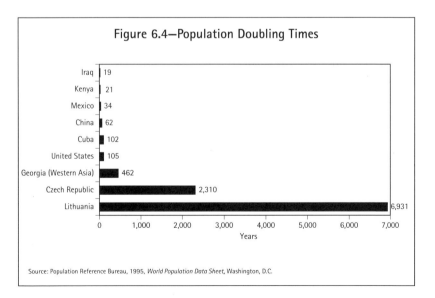

Figure 6.4—Population Doubling Times

Country	Years
Iraq	19
Kenya	21
Mexico	34
China	62
Cuba	102
United States	105
Georgia (Western Asia)	462
Czech Republic	2,310
Lithuania	6,931

Source: Population Reference Bureau, 1995, *World Population Data Sheet*, Washington, D.C.

the chances of starvation are lessened. A 6-year-old child in a poor household may take care of siblings, graze animals, and gather water and fuelwood. Poor women also have less access to birth control and less ability to pay for it. Education is correlated with decreased birth rates, but poor women tend to be uneducated. And without access to employment, women have few options. In some cases, children are also a security in old age. In India, for example, 80 percent of elderly people live with their children.[12]

Sometimes religious beliefs contribute to the decision to have many children or not to limit family size. Ironically, many traditional religious beliefs were developed at a time when the world had less than 3 percent of its current population. Those traditional beliefs are now coming into conflict with the modern reality of rapid population growth.

Consequences

The consequences of human population growth are many. In some countries, populations have already exceeded the ability of the environment to sustain them. For example, Ethiopians no longer have enough productive land to feed themselves. When population densities were low, farmers moved to new areas and the land could recover. Now there is nowhere else to go.

Land in many countries has been deforested, planted with crops, or overgrazed to the point that it has turned to desert. Precious topsoil is eroded in the wake of deforestation and overuse. Twenty-five percent of agricultural land is affected by human-caused soil degradation.[13] Groundwater supplies are exhausted or polluted. People who depend on wood for fuel have to travel farther as nearby forests are shared by an increasing number of people. People who use oil and gasoline for fuel are adding greenhouse gases to the atmosphere. More than a billion people breathe air with excessive amounts of particulates.[14] Life expectancies are actually going down in some areas of eastern Europe because pollution has penetrated the air, the land, and the water.

As people take over more land, there is less space for wild animals and plants, which together constitute the very support system that regulates our weather, pollinates our crops, cleans our water, and controls potential pests. In developing countries, the poverty that pushes overuse of natural resources causes the environment to suffer. In turn, degraded land produces less food, less clean drinking water, and less fuelwood, all of which are so desperately needed. Environmental problems, such as expanding deserts, deforestation, and soil erosion, force many people to flee their homes. These environmental refugees are thought to make up 44 percent of the world's refugees.[15]

It's easy to blame global environmental problems on the expanding populations of the developing world, but more important than growth rate is the way people live. Governments of poor countries often overuse natural resources to earn money to pay enormous debts. Hungry citizens overexploit the environment just to survive. But, it's the developed world that is disproportionately responsible for environmental problems. For instance, citizens of the developed world make up only one-fifth of the world's population, yet they are responsible for two-thirds of the atmospheric greenhouse gases.[16] Americans consume 24 percent of the world's energy, but make up only 5 percent of the world's population.[17]

Residents of the United States may be experiencing population pressure from growing numbers and from our consumption habits. The growth in Florida's population alone requires that each day highways need to be expanded by two miles, two new classrooms with two new teachers need to be added, 111,108 additional gallons of water are used, 94,560 additional gallons of waste water are created, and 3,546 additional pounds of solid waste need to be treated.[18] Turning around environmental problems is not simply a matter of controlling the growth rate in developing countries; it also means modifying the lifestyles of the people living in the developed world.

Slowing Population Growth

International family planning programs have not been as successful as they might have been for a variety of reasons. Long trips, long waits at clinics, and a limited choice of family planning methods are deterrents to family planning. Studies have shown that birth control, which costs more than 1 percent of the family's annual income, is also a deterrent. In addition, cultural values and communication problems, such as opposition from husbands, fear of being examined, and religious taboos, prevent women from practicing birth control. Family planning assistance from developed countries has also become sporadic as money gets tighter and as some governments have come under increasing pressure from right-to-life groups.

There are two major strategies that have been successful in lowering population growth rates. One is economic development. Countries that are economically stable tend to have more educational opportunities, health care, jobs, and social equality. Secondary education for women in particular is effective since more highly educated women tend to delay childbearing, are more receptive to new ideas such as family planning, and have more work opportunities.

The second effective strategy is family planning. Access to birth control that is inexpensive, easily obtainable, affordable, and compatible with the culture's beliefs and realities can slow the birth rate.

There are a few signs that population growth is slowing. Since the early 1960s, worldwide contraceptive use has gone from 10 percent of couples to 50 percent of couples.[19] Worldwide, fertility has dropped by one child per women in the last 20 years.[20] But even with these successes, stopping the exponential growth in population will require these types of changes to happen much faster.

The alternatives to these strategies are not pleasant. The consequences of global warming, ozone depletion, deforestation, extinction, and pollution are all capable of diminishing our numbers more brutally.

Facts About Global Population Growth

General

In 1993, the world population was 5.57 billion people. It took thousands of years to reach 1 billion. It took 100 years to reach 2 billion and only 30 years to add another billion. To increase from 3 billion to 4 billion took only 12 years.[21]

The population in mid-1995 was 5.7 billion.

The United Nations projects the world population will be 8.5 billion by 2025 and 10 billion by 2050. Ninety percent of the projected population growth will come from developing countries.[22]

More than half of the world's population is below reproductive age.[23]

The world's population is increasing by 3 people per second, a 250,000 every day, and up to 100 million a year.[24]

From 1750 to 1930, the global population grew most rapidly in what is now the developed world. Since 1930, the population has grown more rapidly in developing countries.[25]

In 1900, only 65 percent of the world's population lived in the Third World. By the end of this century, 78 percent of the world's population will be in the developing world.[26]

Projected deaths from AIDS for the 1990s are equal to 1 month of global population growth.[27]

Fourteen million children under 5 years old die every year.[28]

An African woman is 500 times more likely to die in childbirth than a woman in the developed world.[29]

Population doubling times:
Kenya—21 years
Brazil—41 years
India—36 years
China—62 years
United States—105[30]

Forty-five percent of the world's population lives in cities.[31]

By 2015, developing countries as a whole will be more than 50 percent urban.[32]

The U.S. population is growing by approximately 2.7 million people per year from birth and immigration. That's equivalent to adding the population of Philadelphia or Washington, D.C., each year.[33]

By 2050 the population of the United States is projected to increase by 133 million, which is like adding thirty-eight cities the size of Los Angeles.[34]

Life expectancy increases:

	1950–1955	1985–1990
China	40.8 years	69.4 years
India	38.7 years	57.9 years
Indonesia	37.5 years	60.2 years
Developed countries	66 years	74 years[35]

Impact on the Environment

With only 20 percent of the world's population, the developed world contributes two-thirds of the greenhouse gases.[36]

The average U.S. citizen uses seventy times more water each year than the average citizen of Ghana.[37]

The average American uses as much energy as 3 Japanese, 6 Mexicans, 12 Chinese, 33 Indians, 147 Bangladeshis, 281 Tanzanians, or 422 Ethiopians.[37]

Three-fourths of people in the developing world make their living from agriculture.[39]

Environmental impacts:
    Desertification—850 million people affected
    Fuelwood shortages—100 million people affected
    Soil erosion and deforestation—500 million people affected
    Excessive particulates in air—1 billion people affected[40]

Up to 44 percent of the world's refugees are environmental refugees.[41]

Slowing Population Growth

Since the early 1960s, contraceptive use has gone from 10 percent of couples to 50 percent of couples.[42]

Fertility has fallen by at least one child per woman in the last 20 years.[43]

Austria, Belgium, and Italy have reached zero population growth. Germany's and Hungary's populations are shrinking.[44]

Population success stories:
    Brazil—Projections for population growth have been modified downward largely due to contraceptive use, urbanization, and education. Projections developed in the 1970s estimated the population would be 212 million by the year 2000. Now experts are projecting 172 million by the same year.[45]

    Egypt—The average number of children per couple dropped from 4.4 in 1985 to 2.9 in 1995. Forty-eight percent of couples of childbearing age use contraceptives.[46]

Kenya—A government campaign to distribute contraceptives has brought down the average number of children per woman from 8 children in the mid-1980s to 5.4 children. Formerly Kenya had the world's highest growth rate.[47]

Global Population Growth Summary

Causes
increasing life expectancy because of:
    expansion of food production
    decline in starvation and malnutrition
    improvements in sanitation
    modern medicine

Consequences
potential for all environmental problems to worsen (increased deforestation, pollution, resource consumption, habitat loss, ozone depletion, extinction, climate change, starvation, and disease)

Appendix: Organizations

Multi-Issue Organizations

Center for Environmental Information, Inc.
50 West Main Street
Rochester, NY 14614
(716) 262-2870

Global Tomorrow Coalition, Inc.
1325 G Street NW
Suite 1010
Washington, DC 20005-3104
(202) 628-4016

Union of Concerned Scientists
26 Church Street
Cambridge, MA 02238
(617) 547-5552

United Nations Environment Programme
Regional North American Office
United Nations Room DC2-0803
New York, NY 10017
(212) 963-8138

World Resources Institute
1709 New York Avenue NW
Washington, DC 20006
(202) 638-6300

Worldwatch Institute
1776 Massachusetts Avenue NW
Washington, DC 20036-1904
(202) 452-1999

Climate Change

Carbon Dioxide Information Analysis Center
Oak Ridge National Laboratory
P.O. Box 2008 (MS-6335)
Oak Ridge, TN 37831-6335
(615) 574-3645

Climate Institute
324 4th Street NW
Washington, DC 20002
(202) 547-0104

Department of Energy
Forrestal Building
1000 Independence Avenue SW
Washington, DC 20585
(202) 586-5000

Environmental Defense Fund
257 Park Avenue South
New York, NY 10010
(212) 505-2100

Greenhouse Action
P.O. Box 16743
Seattle, WA 98116-0743
(206) 937-4346

Greenhouse Crisis Foundation
1130 17th Street NW
Suite 630
Washington, DC 20036
(202) 466-2823

International Institute for Energy Conservation
750 1st Street NE
Washington, DC 20002
(202) 842-3388

National Oceanic and Atmospheric Administration
Environmental Research Laboratories
Public Affairs Office
Boulder, CO 80303
(303) 497-6286

Natural Resources Defense Council
40 West 20th Street
New York, NY 10011
(212) 727-2700

Renew America
1400 16th Street NW
Suite 710
Washington, DC 20036
(202) 232-2252

Ozone Depletion

Greenhouse Crisis Foundation
1130 17th Street NW
Suite 630
Washington, DC 20036
(202) 466-2823

National Oceanic and Atmospheric Administration
Environmental Research Laboratories
Public Affairs Office
Boulder, CO 80303
(303) 497-6286

Tropical Deforestation

American Forests
1516 P Street NW
Washington, DC 20005
(202) 667-3300

Rainforest Action Network
450 Sansome, Suite 700
San Francisco, CA 94111
(415) 398-4404

Rainforest Alliance
65 Bleecker Street
New York, NY 10012
(212) 677-1900

U.S. Department of Agriculture
Forest Service
P.O. Box 96090
Washington, DC 20090-6090
(202) 205-0957

World Forestry Center
4033 SW Canyon Road
Portland, OR 97221
(503) 228-1367

World Wildlife Fund
1250 24th Street NW
Washington, DC 20037
(202) 293-4800

Biodiversity and Extinction

Conservation International
1015 18th Street NW
Suite 1000
Washington, DC 20036
(202) 429-5660

Defenders of Wildlife
1101 14th Street NW
Suite 1400
Washington, DC 20005
(202) 682-9400

Appendix: Organizations

Department of the Interior
Interior Building
1849 C Street NW
Washington, DC 20240
(202) 208-1100

Endangered Species Coalition
666 Pennsylvania Avenue SE
Washington, DC 20003
(202) 547-9009

Fish and Wildlife Reference Service
5430 Grosvenor Lane
Bethesda, MD 20814
(301) 492-6403

National Audubon Society
700 Broadway
New York, NY 10003-9501
(212) 979-3000

National Wildlife Federation
1400 16th Street NW
Washington, DC 20036-2266
(202) 797-6800

The Nature Conservancy
1815 North Lynn Street
Arlington, VA 22209
(703) 841-5300

Wildlife Conservation Society
185th Street and Southern Boulevard
Bronx, NY 10460-1099
(718) 220-1099

World Wildlife Fund
1250 24th Street NW
Washington, DC 20037
(202) 239-4800

Pollution

Acid Rain Foundation
1410 Varsity Drive
Raleigh, NC 27606
(919) 828-9442

Air and Waste Management Association
3 Gateway Center
Four West
Pittsburgh, PA 15222
(412) 232-3444

American Water Resources Association
950 Herndon Parkway, Suite 300
Herndon, VA 22070-5528
(703) 904-1225

Center for Marine Conservation
1725 DeSales Street NW
Suite 500
Washington, DC 20036
(202) 429-5609

Clean Water Fund
National Office
1320 18th Street NW
Washington, DC 20036
(202) 457-0336

Environmental Protection Agency
401 M Street SW
Washington, DC 20460
(202) 260-2090

Freshwater Foundation
Spring Hill Center
725 County Road 6
Wayzata, MN 55391
(612) 449-0092

The Groundwater Foundation
P.O. Box 22558
Lincoln, NE 68542-2558
(402) 434-2740

Inform, Inc.
120 Wall Street
New York, NY 10005
(212) 361-2400

International Institute for Energy Conservation
750 1st Street NE
Washington, DC 20002
(202) 842-3388

The National Ground Water Association
6375 Riverside Drive
Dublin, OH 43017
(614) 761-1711

Global Population Growth

Population Crisis Committee
1120 19th Street NW, Suite 550
Washington, DC 20036
(202) 659-1833

The Population Institute
110 Maryland Avenue NE
Washington, DC 20002
(202) 544-3300

Population Reference Bureau, Inc.
1875 Connecticut Avenue NW
Suite 520
Washington, DC 20009
(202) 483-1100

Zero Population Growth, Inc.
1400 16th Street, NW
Suite 320
Washington, DC 20036
(202) 332-2200

Notes

Chapter 1: Climate Change

1. J. Leggett, ed., 1990, *Global Warming: The Greenpeace Report*, Oxford University Press, New York, NY.
2. G. Lean and D. Hinrichsen, 1992, *Atlas of the Environment*, Harper Perennial, New York, NY.
3. D. E. Abrahamson, ed., 1989, *The Challenge of Global Warming*, Island Press, Washington, DC, and Covelo, CA.
4. D. E. Fisher, 1990, *Fire and Ice: The Greenhouse Effect, Ozone Depletion, and Nuclear Winter*, Harper and Row, New York, NY.
5. D. E. Abrahamson, ed., 1989.
6. Ibid.
7. D. E. Fisher, 1990.
8. D. E. Abrahamson, ed., 1989.
9. R. C. Cowen, *Christian Science Monitor*, June 1, 1994.
10. B. McKibben, 1989, *The End of Nature*, Random House, New York, NY.
11. D. E. Abrahamson, ed., 1989.
12. Ibid.
13. Ibid.
14. Beaumont/Ghazi, *London Observer*, September 5, 1994.
15. D. E. Fisher, 1990.
16. G. Lean and D. Hinrichsen, 1992.
17. J. Leggett, ed., 1990.
18. D. E. Fisher, 1990.
19. D. E. Abrahamson, ed., 1989.
20. Reuters News Service, *The Washington Post*, September 18, 1994.
21. M. Crenson, *The Dallas Morning News*, August 15, 1994.
22. P. J. Cohn, "Gauging the Biological Impacts of the Greenhouse Effect: How Might Species Cope with a Warmer World?", *Bioscience*, March 1989, vol. 39, pp. 142–146.
23. G. Lean and D. Hinrichsen, 1992.
24. D. E. Abrahamson, ed., 1989.

Chapter 2: Ozone Depletion

1. P. Power, Jr., "The Methyl Bromide Ban Years Away: Farmers Readying Now," *The Tampa Tribune*, September 7, 1994.
2. G. Lean and D. Hinrichsen, 1992, *Atlas of the Environment*, HarperPerennial, New York, NY.
3. J. Gliedman, "Is the Pact Too Little Too Late?", *The Nation*, October 10, 1987, vol. 245, pp. 376–380.
4. K. Doyle, "Ozone Alert," *E: The Environmental Magazine*, August 1994, vol. V, pp. 22–24.
5. G. Lean and D. Hinrichsen, 1992.
6. K. Doyle, 1994, "Ozone Alert."
7. T. Noah, *The Wall Street Journal*, January 1, 1995.
8. J. Gliedman, 1987.
9. P. Poore and B. O'Donnell, "Ozone," *Garbage*, September/October 1993, vol. V, pp. 24–29.
10. G. Lean and D. Hinrichsen, 1992.
11. K. Doyle, 1994.
12. Ibid.
13. Ibid.

Chapter 3: Tropical Deforestation

1. World Resources Institute, 1994, *World Resources 1994–95: A Guide to the Global Environment*, Oxford University Press, New York, NY.
2. N. Myers, 1984, *The Primary Source*, W. W. Norton, New York, NY.
3. G. Lean and D. Hinrichsen, 1992, *Atlas of the Environment*, HarperPerennial, New York, NY.
4. N. Myers, 1984.
5. World Resources Institute, 1994.
6. G. Lean and D. Hinrichsen, 1992.
7. Ibid.
8. World Resources Institute, 1994.
9. G. Lean and D. Hinrichsen, 1992.
10. World Resources Institute, 1994.
11. Ibid.
12. G. Lean and D. Hinrichsen, 1992.
13. Ibid.
14. World Resources Institute, 1994.

15. Anonymous, *Greenwire*, May 11, 1995.
16. Ibid.
17. G. Lean and D. Hinrichsen, 1992.
18. S. Hecht and A. Cockburn, 1990, *The Fate of the Forest: Developers, Destroyers, and Defenders of the Amazon*, HarperPerennial, New York, NY.
19. T. Ross, *San Francisco Examiner*, March 20, 1995.
20. N. Myers, 1984.
21. World Resources Institute, 1994.
22. Ibid.
23. Ibid.
24. Ibid.
25. S. Hecht and A. Cockburn, 1990.
26. Ibid.
27. T. Ross, 1995.
28. W. Mwangi, *Nairobi Nation*, September 23, 1994.
29. G. Lean and D. Hinrichsen, 1992.
30. Ibid.
31. World Resources Institute, 1994.
32. G. Lean and D. Hinrichsen, 1992.
33. Ibid.
34. Ibid.
35. S. Hecht and A. Cockburn, 1990.
36. Ibid.
37. Ibid.
38. G. Lean and D. Hinrichsen, 1992.
39. Ibid.

Chapter 4: Biodiversity and Extinction

1. World Resources Institute, 1994, *World Resources 1994-95: A Guide to the Global Environment*, Oxford University Press, New York, NY.
2. E. O. Wilson, ed., 1988, *Biodiversity*, National Academy Press, Washington, DC.
3. E. O. Wilson, 1992, *The Diversity of Life*, W. W. Norton, New York, NY.
4. E. O. Wilson, ed., 1988.
5. T. Hilchey, "New Kangaroo Species Reported," *The Washington Post*, July 25, 1994.

6. World Resources Institute, 1994.
7. W. Reid, "Status and Trends in U.S. Biodiversity," in *Different Drummer*, Summer 1994, Cascade Holistic Economic Consultants, Oak Grove, OR.
8. E. O. Wilson, ed., 1988.
9. *Global Biodiversity Strategy*, World Resources Institute, The World Conservation Union, and United Nations Environment Programme, 1992.
10. World Resources Institute, 1994.
11. E. O. Wilson, ed., 1988.
12. *Global Biodiversity Strategy*, 1992.
13. *Global Biodiversity Strategy*, 1992.
14. N. Myers, 1979, *The Sinking Ark: A New Look at the Problem of Disappearing Species*, Pergamon Press, New York, NY.
15. World Resources Institute, 1994.
16. G. Lean and D. Hinrichsen, 1992, *Atlas of the Environment*, HarperPerennial, New York, NY.
17. World Resources Institute, 1994.
18. G. Lean, D. Hinrichsen, and A. Markham, 1990, *Atlas of the Environment*, Prentice-Hall, New York, NY.
19. E. O. Wilson, ed., 1988.
20. World Resources Institute, 1994.
21. G. Lean, D. Hinrichsen, and A. Markham, 1990.
22. E. C. Wolf, 1987, "On the Brink of Extinction: Conserving the Diversity of Life," Worldwatch paper 78, Worldwatch Institute, Washington, DC.
23. G. Lean and D. Hinrichsen, 1992.
24. Ibid.
25. E. O. Wilson, ed., 1988.
26. Ibid.
27. World Resources Institute, 1994.
28. E. O. Wilson, ed., 1988.
29. W. Reid, 1994.
30. G. Lean and D. Hinrichsen, 1992.
31. N. Myers, 1979.
32. G. Lean and D. Hinrichsen, 1992.
33. W. Reid, 1994.
34. E. O. Wilson, ed., 1988.
35. W. Reid, 1994.
36. E. O. Wilson, ed., 1988.

37. Ibid.
38. P. Ehrlich and A. Ehrlich, 1981, *Extinction: The Causes and Consequences of the Disappearance of Species*, Random House, New York, NY.
39. E. O. Wilson, ed., 1988.
40. Ibid.
41. G. Lean, D. Hinrichsen, and A. Markham, 1990.
42. Ibid.
43. E. O. Wilson, ed., 1988.
44. T. Hilchey, 1994.
45. E. O. Wilson, ed., 1988.
46. N. Myers, 1979.
47. World Resources Institute, 1994.
48. J. A. McNeely, K. R. Miller, W. V. Reid., R. A. Mittermeier, and T. B. Werner, 1990, *Conserving the World's Biological Diversity*, International Union for the Conservation of Nature and Natural Resources, World Resources Institute, Conservation International, World Wildlife Fund–U.S., and the World Bank.
49. E. O. Wilson, ed., 1988.
50. N. Myers, 1979.
51. E. C. Wolf, 1987.
52. W. Reid, 1994.
53. G. Lean, D. Hinrichsen, and A. Markham, 1990.
54. World Resources Institute, 1994, *The 1994 Information Please Environmental Almanac*, Houghton Mifflin Company, New York, NY.
55. TRAFFIC Wildlife Trade Fact Sheets, 1995.
56. G. Lean and D. Hinrichsen, 1992.
57. World Resources Institute, 1994, *The 1994 Information Please Environmental Almanac*.
58. G. Lean and D. Hinrichsen, 1992.
59. Ibid.
60. Ibid.

Chapter 5: Pollution

1. World Resources Institute, 1990, *World Resources 1990–91: A Guide to the Global Environment*, Oxford University Press, New York, NY.
2. Ibid.
3. B. Lawren, 1990, "Plastic Rapt," *National Wildlife*, October/November, vol. 28, pp. 10–16.

4. T. Wagner, 1994, *In Our Backyard: A Guide to Understanding Pollution and Its Effects*, Van Nostrand Reinhold, New York, NY.
5. Ibid.
6. Ibid.
7. Ibid.
8. G. Lean and D. Hinrichsen, 1992, *Atlas of the Environment*, HarperPerennial, New York, NY.
9. Ibid.
10. G. Lean, D. Hinrichsen, and A. Markham, 1990, *Atlas of the Environment*, Prentice-Hall, New York, NY.
11. T. Wagner, 1994.
12. Ibid.
13. Ibid.
14. G. Lean and D. Hinrichsen, 1992.
15. T. Wagner, 1994.
16. G. Lean and D. Hinrichsen, 1992.
17. J. Millman, "Exporting Hazardous Wastes," *Technology Review*, April 1989, vol. 92, p. 6.
18. T. Land, "Managing Toxic Waste," *The New Leader*, November 27, 1989, vol. 72, p. 4.
19. G. Lean and D. Hinrichsen, 1992.
20. Ibid.
21. G. Lean, D. Hinrichsen, and A. Markham, 1990.
22. G. Lean and D. Hinrichsen, 1992.
23. Ibid.
24. J. Gallagher, *Chicago Tribune*, September 7, 1994.
25. T. Fitzpatrick, "Oh Say Can You See?: Haze Dims America's Vistas," *The Futurist*, March/April 1991, vol. 25, pp. 25–27.
26. Environmental Protection Agency, *Air Quality Trends*, September 1995.
27. K. Ingley, *Phoenix Arizona Republic*, December 30, 1994.
28. Anonymous, *Frankfurter Rundschau*, September 26, 1995.
29. M. Treshow, 1971, *Whatever Happened to Fresh Air?*, University of Utah Press, Salt Lake City, UT.
30. D. R. Blake and F. S. Rowland, "Urban Leakage of Liquified Petroleum Gas and Its Impact on Mexico City Air Quality," *Science*, August 18, 1995, vol. 269, pp. 953–956.
31. D. Elsom, 1987, *Atmospheric Pollution: Causes, Effects, and Control Policies*, Basil Blackwell, New York, NY.

32. C. Benedict, 1991, "Dirty Germany," *Buzzworm*, March/April 1991, vol. 3, pp. 30–45.
33. T. Wagner, 1994.
34. G. Lean, D. Hinrichsen, and A. Markham, 1990.
35. "1991 Global Report," 1991, *Buzzworm*, January/February, vol. 3, pp. 30–45.
36. T. Wagner, 1994.
37. C. Williams, 1995, *Los Angeles Times*, April 27, 1995.
38. G. Lean and D. Hinrichsen, 1992.
39. Ibid.
40. World Resources Institute, 1990.
41. G. Lean, D. Hinrichsen, and A. Markham, 1990.
42. T. Wagner, 1994.
43. Anonymous, *Chicago Tribune*, January 6, 1995.
44. Associated Press, June 6, 1995.
45. "1991 Global Report," 1991.
46. World Resources Institute, 1990.
47. Anonymous, *Chicago Tribune*, January 6, 1995.
48. G. Lean and D. Hinrichsen, 1992.
49. Ibid.
50. Ibid.
51. T. Wagner, 1994.
52. G. Lean and D. Hinrichsen, 1992.
53. S. A. Gerlach, 1981, *Marine Pollution: Diagnosis and Therapy*, Springer-Verlag, New York, NY.
54. J. McCaull and J. Crossland, 1974, *Water Pollution*, Harcourt Brace Jovanovich, New York, NY.
55. R. Mackay, *London Observer*, October 10, 1995.
56. World Resources Institute, 1990.
57. R. Mackay, 1995.
58. R. Ritter, *Chicago Sun-Times*, August 3, 1994.
59. T. Wagner, 1994.
60. Ibid.
61. G. Lean and D. Hinrichsen, 1992.
62. World Resources Institute, 1990.
63. Ibid.
64. T. Wagner, 1994.
65. G. Lean, D. Hinrichsen, and A. Markham, 1990.
66. T. Wagner, 1994.

67. G. Lean, D. Hinrichsen, and A. Markham, 1990.
68. World Resources Institute, 1990.
69. T. Fitzpatrick, 1991.
70. Ibid.
71. G. Lean and D. Hinrichsen, 1992.
72. T. Wagner, 1994.
73. Ibid.
74. G. Lean, D. Hinrichsen, and A. Markham, 1990.
75. G. Lean and D. Hinrichsen, 1992.
76. Ibid.
77. World Resources Institute, 1990.
78. T. Wagner, 1994.
79. Environmental Protection Agency, 1995.
80. K. Ingley, 1994.
81. Anonymous, *Frankfurter Rundschau,* September 26, 1995.
82. B. McKibben, 1989, *The End of Nature,* Random House, New York, NY.
83. G. Lean, D. Hinrichsen, and A. Markham, 1990.
84. B. McKibben, 1989.
85. World Resources Institute, 1990.
86. Ibid.
87. Ibid.
88. G. Lean, D. Hinrichsen, and A. Markham, 1990.
89. G. Lean and D. Hinrichsen, 1992.
90. "1991 Global Report," 1991.
91. T. Wagner, 1994.
92. Ibid.
93. G. Lean and D. Hinrichsen, 1992.
94. Ibid.
95. T. Wagner, 1994.
96. S. A. Gerlach, 1981.
97. T. Wagner, 1994.
98. S. A. Gerlach, 1981.
99. Ibid.
100. J. McCaull and J. Crossland, 1974.
101. World Resources Institute, 1990.
102. Anonymous, *Frankfurter Rundschau,* September 26, 1995.

Chapter 6: Global Population Growth

1. Sir F. Graham-Smith, ed., 1995, *Population—The Complex Reality: A Report of the Population Summit of the World's Scientific Academies*, North American Press, Golden, CO.
2. N. Sadik, "World Population Continues to Rise," *The Futurist*, March–April 1991, vol. 25, pp. 9–14.
3. Sir F. Graham-Smith, ed., 1995.
4. Ibid.
5. G. Lean and D. Hinrichsen, 1992, *Atlas of the Environment*, HarperPerennial, New York, NY.
6. J. J. Spengler, 1978, *Facing Zero Population Growth: Reactions and Interpretations, Past and Present*, Duke University Press, Durham, NC.
7. Sir F. Graham-Smith, ed., 1995.
8. V. D. Abernethy, 1993, *Population Politics: The Choices That Shape Our Future*, Insight Books, New York and London.
9. "Zero Population Growth," 1995, Teen PACK, Washington, DC.
10. Ibid.
11. Sir F. Graham-Smith, ed., 1995.
12. Ibid.
13. Ibid.
14. A. B. Durning, 1989, "Poverty and the Environment: Reversing the Downward Spiral," Worldwatch paper 92, Worldwatch Institute, Washington, DC.
15. N. Myers and J. Kent, 1995, *Environmental Exodus: An Emerging Crisis in the Global Arena*, The Climate Institute, Washington, DC.
16. G. Lean, D. Hinrichsen, and A. Markham, 1990, *Atlas of the Environment*, Prentice-Hall, New York, NY.
17. "Zero Population Growth," 1995.
18. L. F. Bouvier and L. Grant, 1994, *How Many Americans? Population, Immigration, and the Environment*, Sierra Club Books, San Francisco, CA.
19. Sir F. Graham-Smith, ed., 1995.
20. World Bank, 1993, *Effective Family Planning Programs*, Washington, DC.
21. Sir F. Graham-Smith, ed., 1995.
22. J. Apler, 1991, "Environmentalists: Ban the (Population) Bomb," *Science*, May 31, 1991, vol. 252, p. 1247.

23. Ibid.
24. N. Sadik, 1991.
25. J. J. Spengler, 1978.
26. Ibid.
27. Sir F. Graham-Smith, ed., 1995.
28. G. Lean, D. Hinrichsen, and A. Markham, 1990.
29. G. Lean and D. Hinrichsen, 1992.
30. Population Reference Bureau, 1995, World Population Reference Sheet, Washington, DC.
31. G. Lean and D. Hinrichsen, 1992.
32. Sir F. Graham-Smith, ed., 1995.
33. "Zero Population Growth," 1995.
34. Ibid.
35. Sir F. Graham-Smith, ed., 1995.
36. G. Lean, D. Hinrichsen, and A. Markham, 1990.
37. Ibid.
38. Zero Population Growth, 1995.
39. A. B. Durning, 1989.
40. Ibid.
41. N. Myers and J. Kent, 1995.
42. Sir F. Graham-Smith, ed., 1995.
43. World Bank, 1993.
44. G. Lean and D. Hinrichsen, 1992.
45. *Financial Times*, September 5, 1994.
46. *Baltimore Sun*, September 4, 1994.
47. *Dallas Morning News*, September 4, 1994.

Glossary

Agroforestry—The intercropping of farm crops and trees.

Albedo effect—The relationship between the "shininess" of the earth's surface and the amount of solar radiation that is reflected back into space.

Background extinction rate—The natural extinction rate over time. The background rate is used to compare the average extinction rate with episodes of accelerated extinctions.

Biodiversity—The variety of life and the ecological functions they perform (functions such as nutrient cycles, pollination, predation, photosynthesis, etc.). Biodiversity is a combination of the words "biological" and "diversity."

Chlorofluorocarbons (CFCs)—A family of chemicals used in refrigeration, air-conditioning, packaging, aerosol propellants, and solvents.

Combustion—The reaction of a substance with oxygen at high temperature which produces heat.

Deciduous—An organism that sheds certain parts regularly. Trees that shed all their leaves during a particular season of each year are deciduous.

Deforestation—A permanent change in land cover. The United Nations defines deforestation as the permanent removal of the crown cover of trees to less than 10 percent of the original amount.

Desertification—Desertlike conditions created when natural forces, such as wind, water, and erosion, are combined with human activities such as agriculture, overgrazing, logging, and deforestation.

DNA—An abbreviation for deoxyribonucleic acid, the basic genetic material found in chromosomes.

Ecosystem—A distinct community of interdependent organisms along with the environment they inhabit and interact with.

Ecosystem diversity—The variety of habitats and ecological processes in the natural world.

Endemic—A species that originated and naturally occurs in only one area.

Eutrophication—The process through which an overabundance of nutrients in a body of water (usually freshwater) can lead to algal blooms, which in turn overwhelm plants and create toxins that kill aquatic animals. Eventually the decomposition of dead organisms deoxygenates the water, killing aquatic animals and plants in greater numbers.

Genetic diversity—The variety of genes within a species.

Greenhouse effect—A normal phenomenon in which transparent gases allow the sun's radiant energy to pass toward the earth while trapping some of the earth's heat energy on the way back to space.

Groundwater—Water that is found in the small spaces in rock and soil between the earth's surface and above a layer of impermeable material.

Hectare—10,000 square meters or 2.471 acres.

Incineration—Disposing of solid, semisolid, liquid, or gaseous combustible materials by burning.

Landfill—Places where waste is buried in the ground.

Mangroves—Specific communities of trees and shrubs that live in the tidal estuaries, salt marshes, and muddy coasts of tropical America and Asia.

Niche—The role an organism plays within an ecosystem, or the specific part of a habitat where an organism lives.

Nonpoint source water pollution—Pollution sources where discharge is difficult to measure, such as agricultural runoff, storm water from cities, and chemicals from road deicing.

Ozone depletion—The destruction of the stratospheric ozone layer caused by particular chlorine- and bromine-containing compounds.

Particulates—Small solid particles or liquid droplets suspended in the air.

Permafrost—Ground that is permanently frozen.

Photosynthesis—The biological synthesis of chemical compounds using light energy.

Point source water pollution—Pollution sources where discharge is measurable, such as industrial plants and sewage pipes.

Pollution—Chemicals, waste, or energy that produces harmful effects when discharged in the environment.

Shaman—A medicine man or healer among certain people.

Species diversity—The variety of living organisms in the world.

Stratosphere—The region of the atmosphere that lies 10 to 30 miles above the earth's surface.

Troposphere—The region of the atmosphere that lies 0 to 10 miles above the earth's surface.

Volatile Organic Compounds (VOCs)—Any carbon compound that is reactive in the atmosphere.

Zero population growth—The point at which the human population is no longer increasing.

References

General

Art, H. W., ed. 1993. *The Dictionary of Ecology and Environmental Science*. New York, NY: Henry Holt. (Definitions of words.)

Brown, L. R. (Updated every year.) *State of the World*. New York, NY: Worldwatch Institute. (A global look at environmental issues. Plenty of data.)

Gore, A. 1992. *Earth in the Balance: Ecology and the Human Spirit*. Boston, MA: Houghton Mifflin. (A captivating account of the many impacts people are having on the planet.)

Goudie, A. 1994. *The Human Impact of the Natural Environment*. Cambridge, MA: The MIT Press. (Discusses human impacts on vegetation, animals, climate, geomorphology, and more.)

Harms, V. 1994. *Almanac of the Environment: The Ecology of Everyday Life*. New York, NY: Grosset/Putman Books. (Ecology book for lay audiences. Simple to read. Many drawings.)

Lean, G., and D. Hinrichsen. 1992. *Atlas of the Environment*. New York, NY: HarperPerennial. (Good explanations of many issues with colorful diagrams and maps.)

McMichael, A. J. 1995. *Planetary Overload: Global Environmental Change and the Health of the Human Species*. New York, NY: Cambridge University Press. (Comprehensive book covering population growth, climate change, ozone depletion, soil and water, biodiversity, urbanization, and impediments to change.)

Meadows, D. H., D. L. Meadows, and J. Randers. 1992. *Beyond the Limits: Confronting Global Collapse, Envisioning a Sustainable Future*. White River Junction, VT: Chelsea Green Publishing Company. (Many charts and graphs. Describes population growth, ozone depletion, technology, and sustainable development.)

World Resources Institute. 1994. *The 1994 Information Please Environmental Almanac*. New York, NY: Houghton Mifflin. (Excellent source of environmental information from around the world.)

World Resources Institute. 1996. *World Resources 1996–97: A Guide to the Global Environment.* New York, NY: Oxford University Press. (Loads of data on global environmental and sustainable development issues.)

Climate Change

Abrahamson, D. E., ed. 1989. *The Challenge of Global Warming.* Washington, DC: Island Press. (More in-depth, but not too technical.)

Fisher, D. E. 1990. *Fire and Ice: The Greenhouse Effect, Ozone Depletion, and Nuclear Winter.* New York, NY: Harper and Row, (A bit technical.)

Leggett, J., ed. 1990. *Global Warming: The Greenpeace Report.* New York, NY: Oxford University Press. (A lot of information, but may be long for the casual reader.)

Schneider, S. H. 1989. *Global Warming: Are We Entering the Greenhouse Century?* New York, NY: Vintage Books. (Clear writing, loaded with information.)

Ozone Depletion

Benedick, R. E., 1991. *Ozone Diplomacy: New Directions in Safeguarding the Planet.* Harvard Cambridge, MA: University Press. (Detailed history of ozone negotiations.)

K. Doyle, 1994. "Ozone Alert," *E: The Environmental Magazine,* August 1994, vol. V, pp. 22–24.

Lemonick, M. D. 1992. "The Ozone Vanishes." *Time,* February 17, 1992, vol. 139, pp. 60–68. (Short and easy to read.)

Litfin, L. T. 1994. *Ozone Discourses: Science and Politics in Global Environmental Cooperation.* New York, NY: Columbia University Press. (Very in-depth coverage of ozone negotiations and the Montreal Protocol.)

Poore, P., and B. O'Donnell. 1993. "Ozone." *Garbage,* September/October 1993, vol. V, pp. 24–29.

Tropical Deforestation

Forsyth, A., and K. Miyata. 1984. *Tropical Nature: Life and Death in the Rainforests of Central and South America,* New York, NY:

Charles Scribner's Sons (Description of life in the tropical rainforest. Getting out of date.)

Hecht, S., and A. Cockburn. 1990. *The Fate of the Forest: Developers, Destroyers, and Defenders of the Amazon*. New York, NY: HarperPerennial. (A good source of information about deforestation in the Amazon.)

Park, C. C. 1992. *Tropical Rainforests*. New York, NY: Routledge. (History, destruction, forest peoples, causes of deforestation, and possible solutions.)

Biodiversity and Extinction

Ehrlich, P., and A. Ehrlich. 1981. *Extinction: The Causes and Consequences of the Disappearance of Species*. New York, NY: Random House. (Good source for explanation of extinction issues, but facts getting out of date.)

Kaufman, L., and K. Mallory, eds. 1993. *The Last Extinction*. Cambridge, MA: The MIT Press. (A variety of extinction-related topics, written by a number of authors.)

Myers, N. 1979. *The Sinking Ark: A New Look at the Problem of Disappearing Species*. New York, NY: Pergamon Press. (Good at explaining issues, but data out of date.)

Wilson, E. O. 1992. *The Diversity of Life*. New York, NY: W. W. Norton. (Excellent reading for lay audiences.)

Wilson, E. O., ed. 1988. *Biodiversity*. Washington, DC: National Academy Press. (Comprehensive. Chapters vary from easily readable to very technical.)

Pollution

Adler, R. W., J. C. Landman, and D. M. Cameron. 1993. *The Clean Water Act 20 Years Later*. Washington, DC: Island Press. (In-depth coverage of water issues. A bit technical.)

Andelman, J. B., and D. W. Underhill. 1987. *Health Effects from Hazardous Waste Sites*. Chelsea, MI: Lewis Publishers. (Written like professional journal articles.)

Elsom, D. 1987. *Atmospheric Pollution: Causes, Effects, and Control Policies*. New York, NY: Basil Blackwell. (A bit technical. Some information may be out of date.)

Erickson, J. 1992. *World Out of Balance: Our Polluted Planet*. Blue Ridge Summit, PA: TAB Books. (Covers pollution in the air,

water, and on land. Some global warming information. A bit technical.)
The League of Women Voters Education Fund. 1993. *The Garbage Primer: A Handbook for Citizens*. New York, NY: Lyons and Burford. (Clear, concise descriptions of resource reduction, recycling, and citizen action. *The Plastic Waste Primer* is also available.)
Newsday. 1989. *Rush to Burn: Solving America's Garbage Crisis?* Washington, DC: Island Press. (Describes garbage disposal problems, incinerators, transporting waste, and possible solutions.)
Symons, J. M. 1995. *Drinking Water: Refreshing Answers to All Your Questions*. College Station, TX: Texas A&M University Press. (Descriptions of health issues, conservation, and household solutions in a question-and-answer format.)
Wagner, T. 1994. *In Our Backyard: A Guide to Understanding Pollution and Its Effects*. New York, NY: Van Nostrand Reinhold. (Comprehensive source on air and water pollution and solid waste.)

Global Population Growth

Brown, L. R., and H. Kane. 1994. *Full House: Reassessing the Earth's Population Carrying Capacity*. New York, NY: W. W. Norton and Company. (Clear writing and good graphics.)
Durning, A. B. 1989. *Poverty and the Environment: Reversing the Downward Spiral*. Worldwatch paper 92. Washington, DC: Worldwatch Institute. (Contact Worldwatch Institute for a copy.)
Ehrlich, P. R., and A. H. Ehrlich. 1988. "Population, Plenty, and Poverty." *National Geographic*, December 1988. Washington, DC. (Concise and readable.)
Ehrlich, P. R., and A. H. Ehrlich. 1990. *The Population Explosion*. New York, NY: Simon and Schuster. (For lay audiences. Written by the authors of *The Population Bomb*.)
Graham-Smith, Sir F., ed. 1995. *Population—The Complex Reality: A Report of the Population Summit of the World's Scientific Academies*. Golden, CO: North American Press. (Series of papers from the Population Summit.)
Hardin, G. 1993. *Living Within the Limits: Ecology, Economics, and Population Taboos*. New York, NY: Oxford University Press. (Covers population and carrying capacity issues.)

Index

A

acid rain, 69, 72–73
Africa. *See also specific countries.*
    carbon dioxide emissions, 5, 16
    climate change, 12
    population growth, 96
    tropical forests and deforestation, 33, 37, 38, 39, 43
agriculture
    biodiversity loss, 52–53, 55
    deforestation, 37–38, 42
    methyl bromide, 21
    pesticides, 64–65
    pollution, 63, 64
    runoff, 79–80
AIDS, 51, 90, 96
Alaska, 10, 13
albedo effect, 2, 35, 41
Amazon, 31, 36, 39, 40, 42, 43, 44
Antarctic, 10, 11, 20, 21, 26, 27, 28
Arctic, 10, 14, 15, 22, 49
Arizona, 66, 81
Arrhenius, Svante, 3
Asia, 33, 37, 38, 43, 44
atmospheric gases. *See* greenhouse gases.
Australia, 4, 12, 15, 23, 29
Austria, 90, 97

B

Bangladesh, 11, 15
Belgium, 90, 97
biodiversity, 47–60
    climate change, 12, 13, 55
    deforestation, 36, 41

    ecosystem diversity, 47
    ecosystem services provided by biodiversity, 50–51
    extinction, 41, 51–56, 58–59
    genetic diversity, 47
    products from, 49–50
    species diversity, 33–34, 42, 43, 47–48, 57
birth control. *See* Family planning.
Borneo, 42
Brazil, 17
    biodiversity, 49, 57
    climate change, 4, 12, 17
    forests, 33, 36, 39, 40, 43, 44
    population growth, 46, 97
Bulgaria, 73, 83

C

cadmium, 71
California, 13, 57, 58, 66, 67, 81, 82
Cambodia, 39
Canada, 12, 13, 15
carbon dioxide
    climate change, 2, 3, 4, 5, 7, 8, 13, 14, 16, 67
    deforestation, releases of carbon dioxide from, 7, 41, 44
carbon monoxide, 62, 66, 67, 69–70, 82, 83
    deforestation, releases of carbon monoxide from, 41, 44
carbon tetrachloride, 20, 21, 24
Caribbean, 22, 33, 37, 42, 43
Central America, 12, 40. *See also specific countries.*

CFCs (chlorofluorocarbons)
    climate change, 2, 3, 7, 17
    ozone depletion, 7, 20, 21, 22,
        23, 24, 26, 27, 28
Chernobyl, 71
China
    biodiversity, 49, 57
    carbon dioxide emissions, 5
    forestry, 39
    ozone depletion, 22, 27
    population growth, 92, 96
Clean Air Act, 66
climate change, 1–18
    ecosystems, effects on, 12–13
    food production, effects on, 11
    health, effects on, 12
    rainfall patterns, 10
    sea level rise, 10–11, 15, 16
    temperature rise, 3, 4, 9, 14
clouds
    climate change, 8, 9, 14
    water cycle in the tropics, 35
coliform bacteria, 74
Colombia, 16, 40, 43, 49, 57
Colorado, 66, 82
consumption, 55, 94
contraceptives. *See* family planning.
coral reefs
    biodiversity of, 48, 58
    bleaching, 12
    ozone depletion, 24
Costa Rica, 33
Czechoslovakia, 73, 82, 83

D
deforestation, 31–46, 97
    climate effects, 4, 8, 41
    definition of, 36
desertification, 41–42, 97
dioxin, 62, 79
dumping. *See* ocean dumping.

E
earth's orbit, 1, 2
ecosystem diversity. *See* biodiversity.

ecosystem services. *See* biodiversity.
Ecuador, 49, 57
Egypt, 11, 97
England, 68
Environmental Protection Agency
    (EPA), 26, 28
Ethiopia, 58, 93
Europe. *See also specific countries.*
    biodiversity loss, 55
    carbon dioxide emissions, 5, 16
    forestry, 38, 39, 40, 44
    ozone depletion, 21, 22, 27
    pollution, 61, 68, 73, 93
eutrophication, 80
extinction, 41, 51–56, 58–59
Exxon *Valdez*, 77, 84

F
family planning, 94–95, 97, 98
fertilizer, 8
Florida, 21, 94
food production. *See* agriculture.
forestry, 38–39, 41, 42, 52
fossil fuels, 4, 6, 9, 69
France, 12
fuelwood, 39, 45, 97

G
general circulation models, 8, 9
genetic diversity. *See* biodiversity.
Germany, 12, 67, 69, 73, 83, 85, 90,
    97
global warming. *See* climate change.
Greece, 12
greenhouse effect, 1, 2, 24. *See also*
    climate change.
greenhouse gases, 94, 97. *See also*
    carbon dioxide, CFCs, and
    methane.
    effects on climate, 1, 2, 15, 69

H
habitat loss, 52
Haiti, 43
Hawaii, 3, 49, 59

hazardous waste, 63–64, 81
heavy metals, 62, 71, 79
Hungary, 83, 90, 97
hunting, 52, 54–55

I

incineration, 62
India, 12, 39, 61, 81, 83, 92, 93, 96
Indonesia
    biodiversity, 47, 49, 57, 58
    forests, 33, 39, 43
    population growth, 92, 96
indoor air pollution, 73–74
introduced species, 53, 59
Italy, 12, 90, 97

J

Japan
    hardwood products, demand for, 38, 39, 44
    ozone depletion, 21, 22, 27
    pollution, 73, 80, 81
    wildlife trade, 55

K

Keeling, Charles D., 3
Kenya, 43, 96, 98

L

landfills, 62
Latin America, 37, 38, 39, 41. *See also specific countries.*
lead, 66, 71, 79, 82
livestock grazing, 6, 7, 40
logging. *See* forestry.

M

Madagascar, 42, 49, 57
Malaysia, 39
Maryland, 17
Mediterranean, 12
mercury, 71, 79

methane
    climate change, 6, 7, 8, 14, 17
    deforestation, 41, 44
    ozone depletion, 20
methyl bromide, 21, 28
methyl chloroform, 20, 26
Mexico, 10, 12, 49, 57, 66, 68, 75, 82
Middle East, 5, 16. *See also specific countries.*
Minnesota, 82
Mississippi River, 13
Montreal Protocol, 7, 24–26, 29
Myanmar, 36, 43

N

Nebraska, 15
Netherlands, 11
Nevada, 66, 81
New York (city), 12, 61, 81, 82
    (state), 67
nitrogen, 62
nitrogen oxides, 20, 66, 69, 82
nitrous oxides, 8
North America. *See also specific countries.*
    biodiversity, 33, 41, 57
    carbon dioxide emissions, 5, 15
    forests, 41
nutrient cycles, 34–35

O

oceans. *See also* pollution.
    climate change, 1, 2, 8, 13
    dumping, 75, 85
    ozone depletion, 24
    sea level rise, 10–11, 15, 16
Ohio, 83
oil, 75–78
overhunting. *See* hunting.
ozone
    depletion, 20, 21, 22, 27
    description of ozone layer, 19, 24, 27

food production impacts of ozone depletion, 23
ground level ozone, 7, 19, 27, 66, 69, 70, 72, 82
health effects of ozone depletion, 23

P

PAN (peroxyacetyl nitrate), 72
PCBs (polychlorinated biphenyls), 70, 79
Papua New Guinea, 39
particulates, 62, 66, 68, 82, 93, 97
Peru, 33, 39, 43, 44, 58
pesticides, 64–65, 84, 85
pH scale, 72–73
Philippines, 3, 39, 61, 81
photochemical smog, 72
Poland, 63, 73, 74, 83
polar ice caps, 1, 10, 11, 15
polar vortex, 21
pollution, 61–87
    air, 66–74, 81–83
    air pollution fallout, 78
    indoor air pollution, 73–74
    land, pollution on, 61–66, 81
    nonpoint source pollution, 74–75
    point source pollution, 74–75
    thermal, 78
    toxic, 70–71
    water, 74–80, 83–85
population growth, 89–98

R

radiation, solar, 1, 2, 21, 22. *See also* ultraviolet radiation.
radioactivity, 71, 78–79, 85
radon, 71, 74
recycling, 81
rice paddies, 6, 7, 8, 14, 17
road building, 37, 40, 53
runoff
    agricultural, 79–80
    urban, 79–80
Russia, 66

S

Sahal, 12
Scandinavia, 13, 73
sea level rise. *See* oceans and climate change.
Siberia, 12, 13
skin cancer, 20, 23, 28
smog, 68. *See also* photochemical smog.
solid waste, 61, 66, 81
South America, 33, 37, 42. *See also specific countries*.
Soviet Union, former, 21, 61
species diversity. *See* biodiversity.
stratosphere, 19, 21, 22, 24, 28
sulfur dioxide, 66, 67, 68–69, 82, 83
sulfur oxides, 62
Sweden, 83

T

Tanzania, 23
termites, 6, 7, 8, 17
Thailand, 38
timber harvesting. *See* forestry.
toxic chemicals, 70
transportation, 13
tropical rainforests, 31–46. *See also* deforestation.
    biodiversity, 48–49, 57
    deforestation, 31–46, 97
    description, 31, 33
    products from, 44–45
troposphere, 19, 22

U

Ukraine, 12, 71
ultraviolet radiation, 19, 20, 23, 24, 28, 29. *See also* radiation, solar.
United Arab Emirates, 4, 15
United States. *See also specific states*.

biodiversity, 41, 49, 52, 53, 55, 57, 58, 59
carbon dioxide emissions, 4, 15
food production, 12
forests and forestry, 37, 38, 39, 44
ozone layer, 20, 21, 22, 23, 27, 28, 29
pollution
   air, 66, 71, 73, 81, 82
   land, 61, 62, 63, 64, 65, 81
   water, 78
population growth, 91, 94, 96, 97
sea level rise, 16
Utah, 66

V
Venezuela, 16, 49, 57
Vermont, 83
Vietnam, 38, 47, 58
VOCs (volatile organic compounds), 70

W
Waldsterben, 41, 73
Washington (DC), 9, 15
   (state), 78, 82
water cycle, 35, 41
water vapor, 2, 8, 9
wildlife trade, 54–55, 59

Z
Zaire, 33, 43, 57

HERNANDO COUNTY
PUBLIC LIBRARY SYSTEM

AUG -- 1998

238 HOWELL AVENUE
BROOKSVILLE FL 34601